T0258109

Diverse Applications of Calorimetry

Diverse Applications of Calorimetry

Edited by **Bruce Rhames**

New York

Published by NY Research Press,
23 West, 55th Street, Suite 816,
New York, NY 10019, USA
www.nyresearchpress.com

Diverse Applications of Calorimetry
Edited by Bruce Rhames

International Standard Book Number: 978-1-63238-117-0 (Hardback)

Printed in the United States of America.

Contents

Preface

Every book is a source of knowledge and this one is no exception. The idea that led to the conceptualization of this book was the fact that the world is advancing rapidly; which makes it crucial to document the progress in every field. I am aware that a lot of data is already available, yet, there is a lot more to learn. Hence, I accepted the responsibility of editing this book and contributing my knowledge to the community.

Calorimetry, as a technique for thermal analysis, has a broad spectrum of applications which are not only restricted to studying the thermal characterization (e.g. melting temperature, denaturation temperature and enthalpy change) of small and large drug molecules, but are also applied to characterization of fuel, metals and oils. Differential Scanning Calorimetry is applied to study the thermal behaviors of drug molecules and excipients by calculating the differential heat flow required to maintain the temperature difference between the sample and reference cells. Microcalorimetry is applied to study the thermal transition and folding of biological macromolecules in dilute solutions and could also be applied in formulation and stabilization of therapeutic proteins. This book consists of research from around the world on the usage of calorimetry on both solid and liquid states of materials and has been grouped under sections: Application of Differential Scanning Calorimetry into Pharmaceuticals, Application of Isothermal Titration Calorimetry for Analysis of Proteins and DNA, and Application of Microcalorimetry to Study Protein Stability and Folding Reversibility.

While editing this book, I had multiple visions for it. Then I finally narrowed down to make every chapter a sole standing text explaining a particular topic, so that they can be used independently. However, the umbrella subject sinews them into a common theme. This makes the book a unique platform of knowledge.

I would like to give the major credit of this book to the experts from every corner of the world, who took the time to share their expertise with us. Also, I owe the completion of this book to the never-ending support of my family, who supported me throughout the project.

Editor

Application of Differential Scanning Calorimetry into Pharmaceuticals

Thermal Stability of the Nanostructured Powder Mixtures Prepared by Mechanical Alloying

Safia Alleg, Saida Souilah and Joan Joseph Suñol

Additional information is available at the end of the chapter

1. Introduction

Nanocrystalline materials present an attractive potential for technological applications and provide an excellent opportunity to study the nature of solid interfaces and to extend knowledge of the structure-property relationship in solid materials down to the nanometer regime. Nanocrystalline materials can be produced by various methods such as mechanical alloying, inert gas condensation, sol–gel process, electrodeposition, chemical vapour deposition, heat treatment of amorphous ribbons, high speed deformation, etc. Mechanical alloying is a non-equilibrium process resulting in solid state alloying beyond the equilibrium solubility limit. During the milling process, mixtures of elemental or prealloyed powders are subjected to heavy plastic deformation through high-energy collision from the balls. The processes of fracturing and cold welding, as well as their kinetics and predominance at any stage, depend mostly on the deformation characteristics of the starting powders. As a result of the induced heavy plastic deformation into the powder particles during the milling process, nanostructured materials are produced by the structural decomposition of coarser-grained structure. This leads to a continuous refinement of the internal structure of the powder particles to nanometer scales.

Solid-state processing is a way to obtain alloys in states far-from-equilibrium. The microstructural manifestations of the departures from equilibrium achieved by mechanical alloying can be classified as follows: (i) **augmented defect concentrations** such as vacancies, interstitials, dislocations, stacking faults, twin boundaries, grain boundaries as well as an increased level of chemical disorder in ordered solid solutions and compounds; (ii) **microstructural refinement** which involves finer scale distributions of different phases and of solutes; (iii) **extended solid solubility**; a stable crystalline phase may be found with solute levels beyond the solubility limit at ambient temperature, or beyond the equilibrium limit at any temperature; and (iv) **metastable phases** which may form during processing like crystalline, quasicrystalline and intermetallic compounds. Chemical reactions can proceed towards equilibrium in stages, and the intermediate stages can yield a metastable phase. In the solid state amorphization reaction, an amorphous alloy can be produced by the reaction of two solid metallic elements. Severe mechanical deformation can lead to metastable states. The deformation forces the production of disturbed configurations or brings different phases into intimate contact promoting solid-state reactions.

The alloying process can be carried out using different apparatus such as planetary mills, attrition mills, vibratory mills, shaker mills, etc. [1]. A broad range of alloys, solid solutions, intermetallics and composites have been prepared in the nanocrystalline, quasicrystalline or amorphous state [2-10]. A significant increase in solubility limit has been reported in many mechanically alloyed systems [11, 12]. Several studies of the alloy formation process during mechanical alloying have led to conflicting conclusions like the interdiffusion of elements, the interactions on interface boundaries and/or the diffusion of solute atoms in the host matrix. Indeed, the alloying process is complex and hence, involves optimization of several parameters to achieve the desired product such as type mill, raw material, milling intensity or milling speed, milling container, milling atmosphere, milling time, temperature of milling, ball-to-powder weight ratio, process control agent, etc. The formation of stable and/or metastable crystalline phases usually competes with the formation of the amorphous phase. For alloys with a negative heat of mixing, the phase formation has been explained by an interdiffusion reaction of the components occurring during the milling process [13]. Even though the number of phases reported to form in different alloy systems is unusually large [14], and property evaluations have been done in only some cases and applications have been explored, the number of investigations devoted to an understanding of the mechanism through which the alloy phase's form is very limited. This chapter summarizes the information available in this area. The obtained disordered structures by mechanical alloying are metastable and therefore, they will experience an ordering transition during heating resulting in exothermic and/or endothermic reactions. The thermal properties of materials are strongly related to the size of nanocrystals essentially when the radius of nanocrystals is smaller than 10 nm. Hence, an important task of thermal analyses is to find the size-dependent function of the thermodynamic amounts of nanocrystalline materials.

2. Thermodynamic stability

The state of a physical system evolves irreversibly towards a time-independent state in which no further macroscopic physical or chemical changes can be seen. This is the state of thermodynamic equilibrium characterized, for example, by a uniform temperature throughout the system but also by other futures. A non-equilibrium state can be defined as a state where irreversible processes drive the system towards the equilibrium state at different rates ranging from extremely fast to extremely slow. In this latter case, the isolated system may appear to have reached equilibrium. Such a system, which fulfils the characteristics of an equilibrium system but is not the true equilibrium state, is called a metastable state. Both stable and metastable states are in internal equilibrium since they can explore their complete phase space, and the thermodynamic properties are equally well defined for metastable states as for stable states. However, only the thermodynamically stable state is in global equilibrium; a metastable state has higher Gibbs energy than the true equilibrium state.

Thermodynamically, a system will be in stable equilibrium, under the given conditions of temperature and pressure, if it is at the lowest value of the Gibbs free energy:

$$G = H - TS \qquad (1)$$

Where H is enthalpy, T absolute temperature and S entropy. According to equation (1), a system can be most stable either by increasing the entropy or decreasing the enthalpy or both. At low temperatures, solids are the most stable phases since they have the strongest atomic bonding (the lowest H), while at high temperatures the -TS term dominate. Therefore, phases with more freedom of atomic movement, such as liquids and gases are most stable. Hence, in the solid-state transformations, a close packed structure is more stable at low temperatures, while a less close packed structure is most stable at higher temperatures. A metastable state is one in internal equilibrium, that is, within the range of configurations to which there is access by continuous change, the system has the lowest possible free energy. However, if there were large fluctuations (the nucleation of a more stable phase), transformation to the new phase would occur if the change in free energy, ΔG, is negative. A phase is non-equilibrium or metastable if it's Gibbs free energy is higher than in the equilibrium state for the given composition. If the Gibbs free energy of this phase is lower than that of other competing phases (or mixtures thereof), then it can exist in a metastable equilibrium. Consequently, non-equilibrium phases can be synthesized and retained at room temperature and pressure when the free energy of the stable phases is raised to a higher level than under equilibrium conditions, but is maintained at a value below those of other competing phases. Also, if the kinetics during synthesis is not fast enough to allow the formation of equilibrium phase(s), then metastable phases could form.

3. Transformation mechanism

During the mechanical alloying process, continuous fracturing, cold welding and rewelding of the powder particles lead to the reduction of grain size down to the nanometer scale, and to the increase of the atomic level strain. In addition, the material is usually under far-from-equilibrium conditions containing metastable crystalline, quasi-crystalline or amorphous phases. All of these effects, either alone or in combination, make the material highly metastable. Therefore, the transformation behaviour of these powders to the equilibrium state by thermal treatments is of both scientific and technological importance. Scientifically, it is instructive to know whether transformations in ball milled materials take place *via* the same transformation paths and mechanisms that occur in stable equilibrium phases or not. Technologically, it will be useful to know the maximal use temperature of the ball milled material without any transformation occurring and thus, losing the special attributes of this powder product. One of the most useful techniques for studying transformation behaviour of metastable phases is differential scanning calorimetry (DSC) or differential thermal analysis (DTA). Hence, a small quantity of the powder milled for a given time is heated at a constant rate to high temperatures under vacuum or in an inert atmosphere to avoid oxidation. Depending on the phase transformations, DSC/DTA scans exhibit endothermic and/or exothermic peaks related to absorption or evolution of heat, respectively, as shown in Fig. 1.

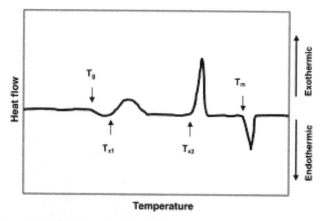

Figure 1. A schematic DSC curve depicting the different stages during crystallization of an amorphous phase where Tg is the glass transition temperature; T_m the melting temperature, T_{x1} and T_{x2} are the onset crystallization temperatures [15].

The values of the peak onset temperature and peak areas depend on the position of the baseline. Therefore, the accurate baseline can be obtained by heating the sample to the desired temperature, then cooled it back to the ambient temperature and then reheated it to higher temperatures. The second DSC scan could be used either as the baseline or subtracted from the first scan to obtain the accurate peak positions and areas. There are two types of transformations: reversible and irreversible. For the former, the product phase will revert back to the parent phase. For example, transformation from one equilibrium phase to another on heating gives rise to an endothermic peak during melting and exothermic peak during cooling. However, during irreversible transformation of metastable phases such as amorphous phases, a peak of the opposite sign is not observed. In fact, there will be no peak at all. Furthermore, because metastable phases are always more energetic than the corresponding equilibrium phases, they often exhibit exothermic peaks in the DSC/DTA curves. If an amorphous alloy powder is heated to higher temperatures, one expects to observe a broad exothermic reaction at relatively low temperatures related to structural relaxation of the amorphous phase, a glass transition temperature as well as one or more exothermic peaks corresponding to crystallization event at higher temperatures. Structural changes that occur during crystallization can be investigated by X-rays diffraction or Mössbauer spectrometry by quenching the sample from a temperature just above the DSC/DTA peak temperature. Transmission electron microscopy investigations can also be conducted to uncover the microstructural and crystal structure changes on a finer scale. In addition, compositional changes can be detected. It may be pointed out, however, that there have not been many detailed crystallization studies of amorphous alloys synthesized by the mechanical alloying process [16].

3.1. Non-isothermal transformation

The crystallization temperature corresponds to the maximum of the exothermic peak, T_p and it increases with increasing heating rate. A relation between heating rate β and position of the transformation peak T_p first described by Kissinger [17], has been extensively used to determine the apparent activation energy for crystallization E_a:

$$\ln\frac{\beta}{T_p^2} = \left(-\frac{E_a}{RT_p}\right) + A \tag{2}$$

Where A is a constant and R is the universal gas constant. The activation energy E_a can be calculated from the slope $\left(\frac{AE_a}{R}\right)$ of the plot $\left(\frac{\beta}{T_p^2}\right)$ against $\left(\frac{1}{T_p}\right)$. Further informations about the transformation temperatures, the number of stages in which the transformation is occurring, details about the product(s) of each individual transformation (crystal structure, microstructure and chemical composition), and the activation energy (and also the atomic mechanism) can be obtained with the combination of DSC/DTA and X-rays diffraction/transmission electron microscopy techniques. The Kissinger method may not be useful in all studies of decomposition. For example, it may not be applicable for metallic glasses which may decompose by nucleation/growth, or a combination of both processes, where the decomposition is seldom described by first-order reaction kinetics [18, 19]. Solid state reactions sometimes exhibit first-order kinetics, this is one form of the Avrami-Erofeev equation (n=1). Such kinetic behaviour may be observed in decompositions of fine powders if particle nucleation occurs on a random basis and growth does not advance beyond the individual crystallite nucleated. The physical interpretation of E_a depends on the details of nucleation and growth mechanisms, and in some cases equation (2) is not valid. For each crystallization peak, the calorimetric results can be explained using the Johnson-Mehl-Avrami-Erofe've kinetics equation [20] for the transformed fraction:

$$\frac{dx}{dt} = K(T)f_n(x) \tag{3}$$

With:

$$f_n(x) = n(1-x)\{-\ln(1-x)\}^{(n-1)/n} \tag{4}$$

$f_n(x)$ gives the transformation rate at time t and temperature T in terms of the rate constant:

$$K(T) = k_0 exp(E/RT) \tag{5}$$

k_0 is the pre-exponential factor; E is the effective activation energy and n is the kinetic exponent. According to the Avrami exponent value, the reaction may be three-dimensional, interface-controlled growth with constant nucleation rate (n=4); three-dimensional, interface-controlled growth with zero nucleation rate (n=3) or diffusion-controlled with growth and segregation at dislocations (n=2/3)[21].

3.2. Isothermal transformation

Isothermal transformation kinetics study at different temperatures can be conducted by the Kolmogorov-Johnson-Mehl-Avrami formalism [22-25] in which the fraction transformed, x, exhibits a time dependence of the form:

$$x(t) = 1 - \exp(-kt)^n \tag{6}$$

Where n is the Avrami exponent that reflects the nucleation rate and/or the growth mechanism; $x(t)$ is the volume of transformed fraction; t is the time, and k is a thermally-activated rate constant. The double logarithmic plot $\ln(-\ln(1 - x))$ against $\ln t$ should give a straight line, the slope of which represents the order of reaction or Avrami parameter n. The rate constant k is a temperature-sensitive factor $k = k_0 exp(AE_a/RT)$, where E_a is the apparent activation energy and k_0 a constant. $x(t)$ corresponds to the ratio between the area under the peak of the isothermal DSC trace, at different times, and the total area. Such analysis was conducted on the phase transformation mechanisms in many mechanically alloyed powders since the milling process occurs at ambient temperature for different milling durations [26-31]. If the Kolmogorov-Johnson-Mehl-Avrami analysis is valid, the value of n should not change with either the volume fraction transformed, V_f or the temperature of transformation. Calka and Radlinski [32] have shown that the usual method of applying the Kolmogorov-Johnson-Mehl-Avrami equation and calculating the mean value of Avrami exponent over a range of volume fraction transformed, may be inappropriate, even misleading, if competing reactions or changes in growth dimensionality occur during the transformation progress. Also, a close examination of the Avrami plots reveals that there are deviations from linearity over the full range of volume fraction transformed [33]. The first derivative of the Avrami plot $\delta\left[\ln(-\ln(1 - n))\right]/\delta \ln t$ against the volume fraction transformed [34], which effectively gives the local value of n with V_f, seems to be more sensitive. Such a plot allows a more detailed evaluation of the data and can emphasize changes in reaction kinetics during the transformation process.

4. Mechanical alloying process

Mechanical alloying has received a great interest in developing different material systems. It is a solid state process that provides a means to overcome the drawback of formation of new alloys starting from mixture of low and/or high melting temperature elements. Mechanical alloying is a ball milling process where a powder mixture placed in the vials is subjected to high-energy collisions from the balls. The two important processes involved in ball milling are fracturing and cold welding of powder particles in a dry high energy ball-mill. The alloying process can be carried out using different apparatus such as planetary or horizontal mills, attrition or spex shaker mill. The elemental or prealloyed powder mixture is charged in the jar (or vial) together with some balls. As a result of the induced heavy plastic deformation into the powder particles during the milling process, nanostructured materials are produced by the structural decomposition of coarser-grained structure. This leads to a continuous refinement of the internal structure of the powder particles down to nanometer scales.

Depending on the microstructure, the mechanical alloying process can be divided into many stages: initial, intermediate, final and complete [35]. Since the powder particles are soft in the early stage of milling, so they are flattened by the compressive forces due to the collisions of the balls. Therefore, both flattened and un-flattened layers of particles come into intimate contact with each other leading to the building up of ingredients. A wide range of particle sizes can be observed due to the difference in ductility of the brittle and ductile powder particles. The relatively hard particles tend to resist the attrition and compressive forces. However, if the powder mixture contains both ductile and brittle particles (Fig. 2a), the hard particles may remain less deformed while the ductile ones tend to bind the hard particles together [10, 36]. Cold welding is expected to be predominant in fcc metals (Fig. 2b) as compared to fracture in bcc and hcp metals (Fig. 2c).

During the intermediate stage of milling, significant changes occur in the morphology of the powder particles. Greater plastic deformation leads to the formation of layered structures (Fig. 2d). Fracturing and cold welding are the dominant milling processes. Depending on the dominant forces, a particle may either become smaller in size through fracturing or may agglomerate by welding as the milling process progresses. Significant refinement in particle size is evident at the final stage of milling. Equilibrium between fracturing and cold welding leads to the homogeneity of the particles at the macroscopic scale as shown in Fig. 2d for the $Fe_{50}Co_{50}$ powder mixture [37, 38]. True alloy with composition similar to the starting constituents is formed at the completion of the mechanical alloying process (Fig. 2e) as evidenced by the energy dispersive X analysis, EDX, (Fig. 2f). The large plastic deformation that takes place during the milling process induces local melting leading to the formation of new alloys through a melting mechanism and/or diffusion at relatively high temperature.

Mechanical alloying is a non-equilibrium process resulting in solid state alloying beyond the equilibrium solubility limit. Several studies of the alloy formation process during mechanical alloying have led to conflicting conclusions such as the interdiffusion of elements, the interactions on interface boundaries and/or the diffusion of solute atoms in the host matrix. Indeed, Moumeni et al. have reported that the FeCo solid solution was formed by the interdiffusion of Fe and Co atoms with a predominance of Co diffusion into the Fe matrix according to the spectrometry results [37]. However, Brüning et al. have shown that the FeCo solid solution was formed by the dissolution of Co atoms in the Fe lattice [39]. Sorescu et al. [40] have attributed the increase of the hyperfine magnetic field to a progressive dissolution of Co atoms in the bcc–Fe phase. Such discrepancies have been attributed to the milling conditions and/or to the fitting procedure of the Mössbauer spectra. The role of grain boundaries, the proportions and the thickness of which are dependent on the milling energy affect thus, the hyperfine structure originating some misinterpretations.

Diffusion in mechanical alloying differs from the steady state diffusion since the balance of atom concentration at the interface between two different components may be destroyed by subsequent fracturing of the powder particles. Consequently, new surfaces with different compositions meet each other to form new diffusion couples when different powder particles are cold welded together. Large difference in composition at the interface therefore promotes interdiffusion. In addition, the change in temperature during the milling process

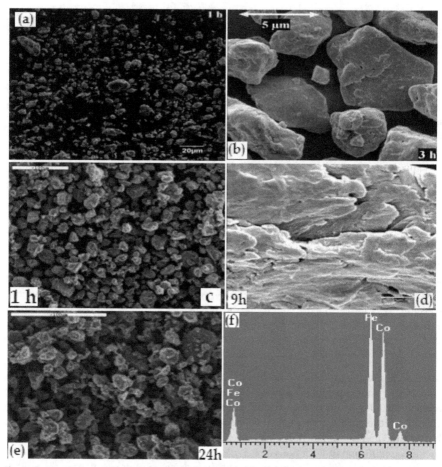

Figure 2. Morphologies of powder particles of the ball-milled $Fe_{75}Si_{15}B_{10}$ (a), $Ni_{20}Co_{80}$ (b), $Fe_{57}Cr_{31}Co_{12}$ (c and d), and $Fe_{50}Co_{50}$ powders (e) with the corresponding EDX analysis (f).

is very significant due to the exothermic reaction causing local combustion. Two major phenomena can contribute to the increase in milling temperature: friction during collisions and localized plastic deformation. At low temperatures, surface diffusion dominates over grain boundary and lattice diffusion. As the temperature is increased, however, grain boundary diffusion predominates, and at higher temperature lattice diffusion becomes the principal mode of diffusion. The first key factor controlling the formation of new alloys is the activation energy which is related to the formation of defects during balls-powder-balls and/or balls-powder-vials collisions. The second key is the vial temperature which is associated with plastic deformation as well as sliding between powder particles and high energetic balls and powder particles. The third key is the crystallite size that is related to the formation of nanometer crystalline structure during the milling process.

5. Experimental section

Mechanical alloying process was used to prepare nanocrystalline and/or amorphous alloys such as Fe, Fe-Co, Fe-Co-Nb-B, Fe-P and Ni-P from pure elemental powders in high-energy planetary ball-mills Fritsch Pulverisette P7 and Retsch PM 400/2, and vibratory ball-mill spex 8000. The milling process was performed at room temperature, under argon atmosphere, with different milling conditions such as rotation speed, ball-to-powder weight ratio, milling time and composition. In order to avoid the temperature increase inside the vials, the milling process was interrupted for 15–30 min after each 30–60 min depending on the raw mixture.

Particles powder morphology evolution during the milling process was followed by scanning electron microscopy. Structural changes were investigated by X-ray diffraction in a $(\theta–2\theta)$ Bragg Brentano geometry with Cu-Kα radiation (λ_{Cu}=0.15406 nm). The microstructural parameters were obtained from the refinement of the X-rays diffraction patterns by using the MAUD program [41, 42] which is based on the Rietveld method. Differential scanning calorimetry was performed under argon atmosphere. Magnetic and hyperfine characterizations were studied by vibrating sample magnetometer and Mössbauer spectrometry, respectively.

6. Fe and FeCo-based alloys

6.1. Fe and Fe-Co powders

Fe and Fe$_{50}$Co$_{50}$ were prepared by mechanical alloying from pure elemental iron and cobalt powders in a planetary ball mill Fritsch P7, under argon atmosphere, using hardened steel vials and balls. The milling intensity was 400 rpm and the ball-to-powder weight ratio was 20:1. A disordered bcc FeCo solid solution is obtained after 24 h of milling (Fig 3), having a lattice parameter, a = 0.2861(5) nm, larger than that of the coarse-grained FeCo phase (a = 0.2825(5) nm). Such a difference in the lattice parameter value may be due to heavily cold worked and plastically deformed state of the powders during the milling process, and to the introduction of several structural defects (vacancies, interstitials, triple defect disorder, etc.).

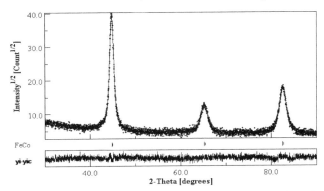

Figure 3. Rietveld refinement of the XRD pattern of the Fe$_{50}$Co$_{50}$ powders milled for 40 h [7].

With increasing milling time, the crystallite size decreases down to the nanometer scale and the internal strain increases. The double logarithmic plot of the crystallite size versus milling time exhibits two-stage behaviour for both Fe and $Fe_{50}Co_{50}$ powders (Fig. 4). A linear fit gives slopes of –0.65 and –0.20 for short and extended milling times, respectively, in the case of Fe; and slopes of –0.85 and –0.03, respectively, for short and extended milling times in the case of $Fe_{50}Co_{50}$ mixture. The critical crystallite size achievable by ball milling is defined by the crossing point between the two regimes with different slopes [43]. Consequently, the obtained critical crystallite sizes are of about 13.8 and 15 nm for Fe and $Fe_{50}Co_{50}$ powders, respectively. By using different milling conditions (mills type, milling intensity and temperature) to prepare nanostructured Fe powders, Börner et al. have obtained the two-regime behaviour, for the grain refinement by using the Spex mill, with slopes of –0.41 and –0.08 for short and extended milling times, respectively. However, the crystallite sizes show only a simple linear relation with slopes of –0.265 and –0.615 by using the Retsch MM2 shaker and the Misuni vibration mill, respectively. The obtained critical crystallite size value was 19 nm [44].

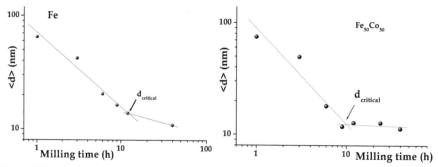

Figure 4. Double logarithmic plot of the crystallite size against milling time for nanostructured Fe and $Fe_{50}Co_{50}$ powders [7].

DSC scans of nanostructured Fe and $Fe_{50}Co_{50}$ powders milled for 40 h are shown in Fig. 5. The non-equilibrium state is revealed by the broad exothermic reaction for both samples, in the temperature range 100–700°C, which is consistent with the energy release during heating due to recovery, grain growth and relaxation processes. As a result of the cold work during the milling process, the main energy contribution is stored in the form of grain boundaries and related strains within the nanostructured grains which are induced through grain boundary stresses [45]. It has been reported that the stored energies during the alloying process largely exceed those resulting from conventional cold working of metals and alloys. Indeed, they can achieve values typical for crystallization enthalpies of metallic glasses corresponding to about 40% of the heat of fusion, ΔH_f [45]. The major sources of mechanical energy storage are both atomic disorder and nanocrystallite boundaries because the transition heats evolving in the atomic reordering and in the grain growth are comparable in value [46].

For the nanostructured Fe powders, the first endothermic peak is linked to the bcc ferro-paramagnetic transition temperature, T_C, and the second peak to the bcc→fcc transition

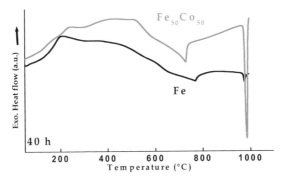

Figure 5. DSC scans of nanostructured Fe and Fe$_{50}$Co$_{50}$ powders milled for 40 h [7].

temperature, $T_{\alpha \to \gamma}$. The depression of Curie temperature with increasing milling duration (Fig. 6), which is ascribed to changes in local order, indicates that the nearest-neighbour coordination is essentially changed in the magnetic nanocrystallites. This reflects to some extent that there are more open disordered spaces or the nearest–neighbour coordination distance in the nanometer sized crystallites is increased, caused by lattice distortion. In fact, if the crystallite sizes are small enough, the structural distortions associated with surfaces or interfaces can lower the Curie temperature. This can be correlated to the increase of the lattice parameter and its deviation from that of the perfect crystal. It has been reported on far-from-equilibrium nanostructured metals, that interfaces present a reduced atomic coordination and a wide distribution of interatomic spacing compared to the crystals and consequently, the atomic arrangement at the grain boundary may be considered close to the amorphous configuration and should therefore alter the Curie temperature. The most reported values of Tc do not deviate strongly from that of the bulk materials. For example, the Tc of 360°C for Ni[C] nanocrystals is in good agreement with that of bulk Ni [47]. Host et al. have reported a Tc value of 1093°C for carbon arc produced Co[C] nanoparticles, in good agreement with the 1115°C value for bulk Co [48]. The Curie temperature of 10 nm Gd is

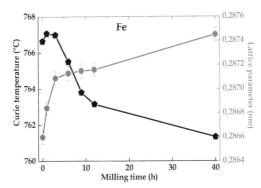

Figure 6. Evolution of the Curie temperature and the lattice parameter of the Fe powders as a function of milling time [7].

decreased by about 10 K from that of coarse-grained Gd while the magnetic transition is broader [49]. According to both Tc and $T_{\alpha \to \gamma}$ temperature values, the paramagnetic nanostructured bcc α–Fe domain is extended by about 50°C at the expense of both magnetic bcc α–Fe and nonmagnetic fcc γ–Fe as compared to coarse-grained bcc α–Fe.

The disorder-order phase transformation temperature of the nanostructured FeCo powders which is nearly constant (~724°C) along of the milling process (Fig. 7), is comparable to that of bulk Fe-Co alloys. It is commonly accepted that Fe-Co undergoes an ordering transition at around 730°C, where the bcc structure takes the ordered α'–CsCl(B2)-type structure [50]. The ordering effect in the FeCo nanocrystals has been revealed by the changes in the magnetization upon heating and the temperature variation of the coercivity on heating and cooling [51]. Also, the phase transformation temperature from bcc–α to fcc–γ structure in the Fe$_{50}$Co$_{50}$ powders is rather milling time independent (~982°C). The lower resistivity of Fe$_{50}$Co$_{50}$ compared to that of pure Fe at 300 K [52] and the higher Curie temperature of Fe$_{50}$Co$_{50}$ suggest that there is less scattering of the conduction electrons by the magnetic excitations. Thus, the Curie temperature cannot be clearly observed because there is a phase transformation from the bcc to fcc form at 985°C.

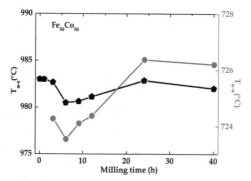

Figure 7. Evolution of the order-disorder, $T_{\alpha \to \alpha'}$, and the bcc→fcc, $T_{\alpha \to \gamma}$, temperatures of the Fe$_{50}$Co$_{50}$ powders as a function of milling time [7].

6.2. Fe-Co-Nb-B powders

Nanostructured and disordered structures obtained by mechanical alloying are usually metastable. Depending on the Nb and B contents, the mechanically alloyed Fe-Co-Nb-B powders structure may be partially amorphous either magnetic and/or paramagnetic. Pure elemental powders of iron (6-8 μm, 99.7%), cobalt (45 μm, 99.8%), niobium (74 μm, 99.85%) and amorphous boron (> 99%) were mixed to give nominal compositions of Fe$_{57}$Co$_{21}$Nb$_7$B$_{15}$ and Fe$_{61}$Co$_{21}$Nb$_3$B$_{15}$ (wt. %), labelled as 7Nb and 3Nb, respectively. The milling process was performed in a planetary ball-mill Fritsch Pulverisette 7, under argon atmosphere, using hardened steel balls and vials. The ball-to-powder weight ratio was about 19/2 and the rotation speed was 700 rpm. For the (Fe$_{50}$Co$_{50}$)$_{62}$Nb$_8$B$_{30}$ mixture, the milling process was performed in a planetary ball-mill Retsch PM400/2, with a ball-to-powder weight ratio of about 8:1 and a rotation speed of 350 rpm. In order to avoid the increase of the temperature inside the vials, the milling process was interrupted after 30 min for 15 min.

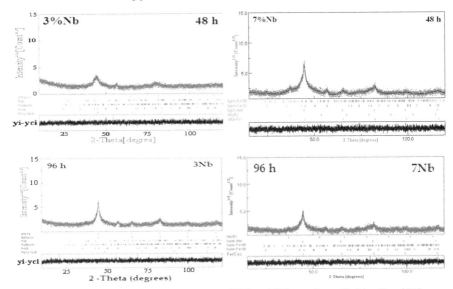

Figure 8. Rietveld refinement of the XRD patterns of 7Nb and 3Nb powders milled for 48 and 96 h [53, 54].

The XRD patterns of 7Nb and 3Nb mixtures milled for 48 h (Fig. 8) are consistent of a large number of overlapping diffraction peaks related to different phases. The Rietveld refinement reveals the formation of a partially amorphous structure of about ~78%, where nanocrystalline tetragonal–Fe_2B, tetragonal–Fe_3B and bcc–FeCo type phases were embedded for 3Nb powders [53]. Whereas, for 7Nb powders, the milling product is a mixture of amorphous (~73.6%), bcc–Nb(B), tetragonal–Fe_2B, orthorhombic–Fe_3B and bcc FeCo type phases [54]. Further milling (up to 96 h) leads to the increase of the amorphous phase proportion for 7Nb and the mechanical recrystallization in the case of 3Nb mixture (Fig. 8) as evidenced by the decrease and the increase of the diffraction peaks intensity, respectively. The formation of the amorphous phase is confirmed by the Mössbauer spectrometry results as shown in Fig. 9. After 48 h of milling, the Mössbauer spectra exhibit more or less sharp absorption lines superimposed upon a broadened spectral component assigned to the structural disorder of the amorphous state [55]. For 3Nb powders, the mechanical recrystallization is confirmed by the emergence of sharp sextet related to the primary crystallization of α–Fe and FeCo after 96 h of milling. However, a stationary state is achieved for 7Nb powders. The increase of the average hyperfine magnetic field, < B_{hyp} >, from 19.18 to 23.14 T after 96 h of milling of 3Nb powders is correlated to the decrease/increase of the amorphous/nanocrystalline relative area. The nanocrystalline (NC) component consists of Fe sites with B_{hyp}>31 T and the interfacial (IF) one is related to the nanostructured Fe–borides with B_{hyp} ranged from 24 to 30 T [28, 56, 57].

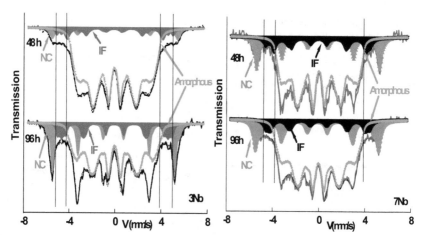

Figure 9. Room temperature Mössbauer spectra of 3Nb and 7Nb powders milled for 48 and 96 h [55].

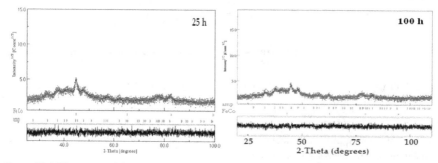

Figure 10. XRD patterns of the $(Fe_{50}Co_{50})_{62}Nb_8B_{30}$ powders milled for 25 and 100 h.

For the $(Fe_{50}Co_{50})_{62}Nb_8B_{30}$ powders mixture milled for 25 and 100 h, the best Rietveld refinements of the XRD patterns were obtained with two components: bcc–FeCo and amorphous phase (Fig. 10). The complete transformation of the heavily deformed FeB and bcc FeCo type phases into an amorphous state is achieved, after 125 h of milling, through the mechanically enhanced solid-state amorphization which requires the existence of chemical disordering, point defects (vacancies, interstitials) and lattice defects (dislocations). Indeed, the severe plastic deformation strongly distorts the unit cell structures making them less crystalline. The powder particles are subjected to continuous defects that lead to a gradual change in the free energy of the crystalline phases above those of amorphous ones, and hence to a disorder in atomic arrangement. The Mössbauer spectra confirm the formation of a paramagnetic amorphous structure, where about 3.8% of FeCo and Fe_2B nanograins are embedded, after 125 h of milling (Fig. 11).

Nanocrystalline $Fe_{72.5}Co_{7.5}Nb_{5+x}B_{15-x}$ with x=0, 5 and 10 at.% labelled as A, B and C, respectively, were prepared by mechanical alloying from pure elemental powders in a planetary ball-mill Retsch PM400, under argon atmosphere, using stainless steel balls and vials. The ball-to-powder weight ratio was about 8:1 and the rotation speed was 200 rpm [58]. The crystallite size decreases with increasing milling duration to about (7.1 ± 0.3) nm for the B-richest alloy (A). The XRD patterns (Fig. 12) reveal the formation of a bcc Fe-rich solid solution after 80 h of milling having an average lattice parameter of about 0.2871 nm for the three alloys.

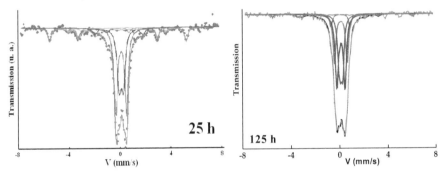

Figure 11. Room temperature Mössbauer spectra of the $(Fe_{50}Co_{50})_{62}Nb_8B_{30}$ powders milled for 25 and 125 h.

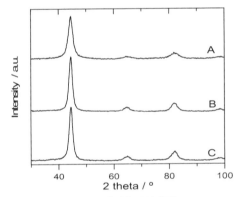

Figure 12. XRD patterns of alloys A, B and C milled for 80 h [58].

Depending on the structural state after each milling time, several exothermic and endothermic peaks appear on heating of the mechanically alloyed Fe-Co-Nb-B powders. Representative DSC scans of 7Nb and 3Nb (Fig. 13) as well as $(Fe_{50}Co_{50})_{62}Nb_8B_{30}$ powder mixtures (Fig. 14) exhibit different thermal effects (Table 1). For all ball milled powders, the first exothermic peak that spreads over the temperature range 100–300°C can be attributed to recovery, strains and structural relaxation. The important heat release (20.56 J/g) for 3Nb powders might be related to the amount of structural defects. The second exothermic peak (2), at 415°C, can be attributed to the α-Fe and/or α-FeCo primary nanocrystallization. This temperature is smaller than that obtained for the ball-milled 7Nb and $(Fe_{50}Co_{50})_{62}Nb_8B_{30}$ powders. Such a difference might be attributed to the Nb content since Co usually increases the onset of crystallization by about 20°C because this atom inhibits atomic movement raising the kinetic barrier for crystallization. The small exothermic peaks centred at ~623.5°C (3) and ~675.7°C (4) in the 3Nb powders can be related to the crystallization of Fe-borides.

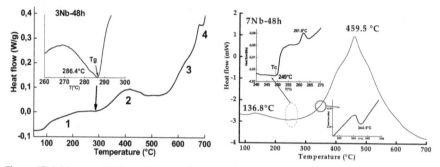

Figure 13. DSC scans of 3Nb and 7Nb powder mixtures milled for 48 h [55].

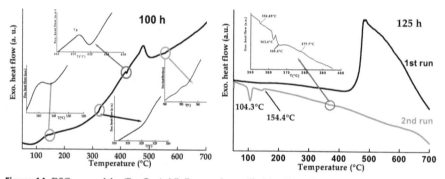

Figure 14. DSC scans of the $(Fe_{50}Co_{50})_{62}Nb_8B_{30}$ powders milled for 100 and 125 h.

Thermal stability of the nanocrystalline phases was investigated by DSC for alloys A, B and C milled for 160 h at a heating rate of 10 K/min (Fig. 15). The broad exothermic process starting at ~400–420 K is due to early surface crystallization (particle surface) and/or internal stress relaxation [58]. In all alloys, an additional exothermic process was detected with a peak temperature between 713 and 743 K. One observes that the peak temperature increases with increasing Nb content from 5 to 15%. This result agrees with those of the ball-milled 3Nb, 7Nb and $(Fe_{50}Co_{50})_{62}Nb_8B_{30}$ mixtures.

The endothermic peak at about 286.4, 344.5 and 420°C for 3Nb, 7Nb and $(Fe_{50}Co_{50})_{62}Nb_8B_{30}$ powders, respectively, that can be attributed to the glass transition temperature, Tg, gives evidence of the amorphous state formation. The glass transition temperature of the $(Fe_{50}Co_{50})_{62}Nb_8B_{30}$ powders increases rapidly up to 25 h of milling, and then remains nearly constant on further milling time (Fig. 16). The increase of Tg might be correlated to the amorphous phase proportion and/or to the change of its composition. The obtained low values compared to those of the amorphous ribbons with the same composition, can be linked to the heterogeneity of the ball-milled samples. The glass transition is not a first order phase transition but a kinetic event dependent on the rearrangement of the system and experimental time scales. Therefore, the transition would be a purely dynamic phenomenon.

Sample Milling time (h)	Peak	T(°C)	ΔH (J/g)	Tg (°C)
$(Fe_{50}Co_{50})_{62}Nb_8B_{30}$ (100 h)	1	138.83	7.58	420
	2	475.76	27.98	
7Nb (48 h)	1	136.8	2.1	344.5
	2	459.5	169.5	
3Nb (48 h)	1	198.5	20.56	286.4
	2	415.0	35.9	
	3	623.5	1.4	
	4	675.7	3.9	

Table 1. Peak temperature, T_P, enthalpy release, ΔH, and glass transition temperature, Tg, of 7Nb and 3Nb powders milled for 48 h, and $(Fe_{50}Co_{50})_{62}Nb_8B_{30}$ mixture milled for 100 h [55].

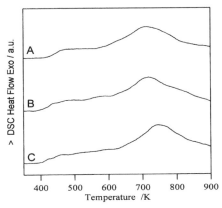

Figure 15. DSC scans at a heating rate of 10 K.min⁻¹ of the ball-milled A, B and C powders for 160 h [58].

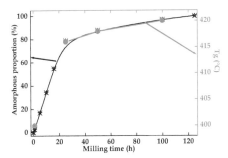

Figure 16. Variation of the glass transition temperature and the amorphous phase proportion of the $(Fe_{50}Co_{50})_{62}Nb_8B_{30}$ powders as a function of milling time.

DSC, which measures heat flow to and from a specimen relative to an inert reference, is the most common thermal analysis method used to measure the glass transition. The heat capacity step change at the glass transition yields three temperature values: onset, midpoint and endset. The midpoint is usually calculated as the peak maximum in the first derivative of heat flow (Fig. 1), although it can be calculated as the midpoint of the extrapolated heat capacities before and after the glass transition. This later is the temperature region where an amorphous material changes from a glassy phase to a rubbery phase upon heating, or *vice versa* if cooling. For example, the glass transition is very important in polymer characterization as the properties of a material are highly dependent on the relationship of the polymer end-use temperature to its Tg. In fact, an elastomer will be brittle if its Tg is too high, and the upper use temperature of a rigid plastic is usually limited by softening at Tg. Therefore, an accurate and precise measure of Tg is a prime concern to many plastics manufacturers and end use designers.

DSC detects the Curie temperature as a change in heat flow and due to the small amount of energy associated with this transition. An endothermic reaction occurs just below the Curie temperature as energy is being absorbed by the sample to induce randomization of the magnetic dipoles. An exothermic event occurs directly after the Curie temperature since no further energy is needed for randomization. Consequently, the line break at about 237°C and 249°C for 3Nb and 7Nb powders (Fig. 13), respectively, can be assigned to the ferro-paramagnetic transition at Curie temperature of the amorphous phase. Those values are comparable to that reported for the amorphous $(Fe_{100-x}Co_x)_{62}Nb_8B_{30}$ bulk metallic glasses [59], where Tc was found to be 245°C for x=0. Accordingly, one can suppose that the amorphous phase composition is Co-free FeBNb-type. Different Tc values of about (157–167)°C and (87–97)°C have been reported for the as-quenched $Fe_{52}Co_{10}Nb_8B_{30}$ and $Fe_{22}Co_{40}Nb_8B_{30}$ alloys [60], respectively. Tc of the residual amorphous phase exhibits antagonist behaviour for both alloys. It decreases with increasing crystalline fraction for the Co–rich $Fe_{22}Co_{40}Nb_8B_{30}$ alloy, and shifts to higher temperature for the Fe–rich $Fe_{52}Co_{10}Nb_8B_{30}$ alloy. Also, lower Tc values in the temperature range (214–230)°C were obtained for the as-cast state and in nanocrystalline $Fe_{77}B_{18}Nb_4Cu$ ribbons annealed at different temperatures [61].

Fig. 17 shows the evolution of Curie temperature of the amorphous phase in the $(Fe_{50}Co_{50})_{62}Nb_8B_{30}$ powders against milling time. Since the amorphous phase Curie temperature is very sensitive to the chemical composition, therefore the progressive decrease of Tc with increasing milling time can be attributed to the increase of B and/or Nb content in the amorphous matrix. It has been reported that the Curie temperature of the FeCoNbB amorphous alloys increases with the B content in the amorphous matrix [62]. Both the first and the second DSC scans of the powders milled for 100 and 125 h, respectively, display many endothermic peaks (see the inset in Fig. 14) that can be attributed to Curie temperatures of different Fe-boride phases and the residual matrix (t=125 h). For example, the endothermic peak at T=579.8°C can be related to the Curie temperature of Fe_3B [63].

The apparent activation energy of the crystallization process in the alloys A, B and C was evaluated by the Kissinger method. The obtained values 2.47±0.07, 2.63±0.05 and 2.71±0.08 eV for alloys A, B and C, respectively, can be associated with grain growth process. The activation energy and the peak temperature variation as a function of Nb content (Fig. 18) reveal that the highest peak temperature and activation energy correspond to the 15%Nb alloy. According to the structural and thermal analysis, it can be concluded that the partial substitution of B by Nb favours the stability of nanocrystalline phase with regard to crystal growth.

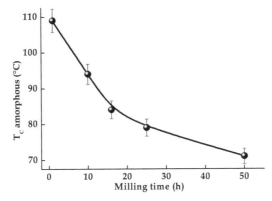

Figure 17. Variation of the amorphous phase Tc in the $(Fe_{50}Co_{50})_{62}Nb_8B_{30}$ powders as a function of milling time.

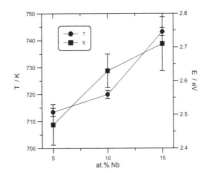

Figure 18. Apparent activation energy and peak temperature of the crystallization process against Nb content for alloys A, B and C milled for 160 h [58].

Stability of the nanostructured Fe-Co-Nb-B powders can be followed by the variation of the magnetic properties such as saturation magnetization, Ms, and coercivity, Hc. The hysteresis loops of ball milled 3Nb powders for 48 h and 7Nb powders for 96 h and heat treated up to 700°C (Fig. 19) display a sigmoidal shape which is usually observed in nanostructured samples with small magnetic domains. This can be correlated to the presence of structural distortions inside grains. One notes that both Ms and Hc values of 3Nb powders are higher than those of 7Nb powders. The increase of Hc from 71 to 115.5 Oe, after heat treatment of the ball milled 3Nb powders for 96 h, points out that the FeCo-rich ferromagnetic grains might be separated by Nb and/or B-rich phase with weaker ferromagnetic properties. Another possible origin for this behaviour is the increase of Fe₂B boride proportion. Nonetheless, for 7Nb mixture Ms increases slightly while Hc remains nearly constant after heat treatment of the powders milled for 48 h. One can conclude that the nanostructured state is maintained after heat treatment.

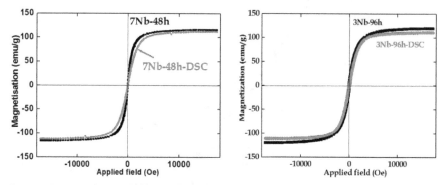

Figure 19. Hysteresis loops of 3Nb and 7Nb powders milled for 96 h and 48 h, respectively, and after heat treatment up to 700°C [55].

7. Ni-P powders

Thermal annealing leads, in general, to the relaxation of the introduced stresses during the milling process. The DSC curves of the ball-milled $Ni_{70}P_{30}$ powders for 3 and 12 h (Fig. 20) display different behaviour on heating at a rate of 10°C.min⁻¹. After the first run up to 700°C (scan a), samples are cooled down to ambient temperature, then reheated in the same conditions. One notes that the DSC signal of the second run (scan b) shows a line without any thermal effect indicating that the phase transformation is achieved during the first run [64]. However, for the first run curve, the enthalpy release spreads over the temperature range (100–650)°C. The large exothermic reactions at temperatures below 300°C can be attributed to recovery and strain relaxation. The DSC curve of the powder milled for 3 h shows a single exothermic peak at 496.4°C. While, after 12 h of milling, the DSC curve reveals several endothermic peaks, and one exothermic peak at 567.6°C. According to the Curie temperature of pure Ni (T_c = 350°C), the endothermic peaks (Fig. 21) can be related to the magnetic transition temperature of dilute Ni(P) solid solutions. However, the exothermic peak might be assigned to a growth process of Ni_2P nanophase. The depression of Tc compared to that of pure Ni indicates that the nearest-neighbour coordinates are essentially changed in the magnetic nanocrystallites by the P additions. The reason for the existence of several magnetic phase states and therefore, several Curie temperatures can be attributed to inhomogeneities since the Curie temperature is sensitive to the chemical short range order and subsequently, to the local Ni environment.

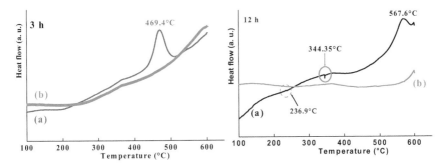

Figure 20. DSC plots of the $Ni_{70}P_{30}$ powders milled for 3 and 12 h at a heating rate of 10°C/min; first (a) and second heating runs (b) [64].

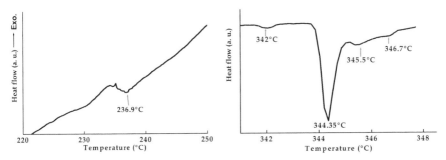

Figure 21. Enlargement of the low temperature regions of the DSC scan of the $Ni_{70}P_{30}$ powders milled for 12 h.

8. Kinetics of powder mixing

8.1. Fe-Mo mixture

The kinetics of Mo dissolution into the α-Fe matrix of the Fe-6Mo mixture has been deduced from the XRD analysis by following the evolution of the (110) diffraction peak intensity of the unmixed Mo as a function of milling time [26]. Since the milling process occurs at room temperature, one can suppose that the temperature is constant. In addition, the milling time can be considered as the necessary time for phase transformation. Consequently, the mixed fraction of Mo which is considered as the fraction transformed, x, can be described by the Johnson-Mehl-Avrami formalism. The double logarithmic plot $\ln(-\ln(1-x))$ versus $\ln t$ leads to the Avrami parameter n, and the rate constant k. Two stages have been distinguished according to the kinetics parameter values: (i) a first stage with $n_1 = 0.83$ and $k_1 = 0.34$; and (ii) a second stage with $n_2 = 0.33$ and $k_2 = 0.73$. The former proves that the Mo dissolution is very slow even non-existent in the early stage of milling (up to 6 h), while the later can be linked to the increased diffusivity by decreasing crystallite size and increasing the grain boundaries area on further milling time.

8.2. $Fe_{27.9}Nb_{2.2}B_{69.9}$ mixture

Amorphization kinetics of the $Fe_{27.9}Nb_{2.2}B_{69.9}$ (at. %) powders has been deduced from the Mössbauer spectrometry results by following the variation of the α-Fe transformed fraction as a function of milling time [27]. The amorphization process can be described by one stage with an Avrami parameter of about n~1 (Fig. 22). This value is comparable to those obtained for transformations controlled by the diffusion at the interface and dislocations segregation with $0.45 < n < 1.1$. This might be correlated to the existence of a high density of dislocations and various types of defects as well as to the crystallite size refinement. Comparable values of the Avrami parameter were obtained for the primary crystallization of the amorphous FeCoNbB alloy prepared by melt spinning [65].

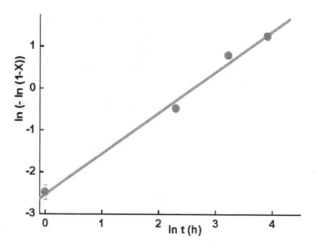

Figure 22. Johnson-Mehl-Avrami plot of the ball-milled $Fe_{27.9}Nb_{2.2}B_{69.9}$ versus milling time [28].

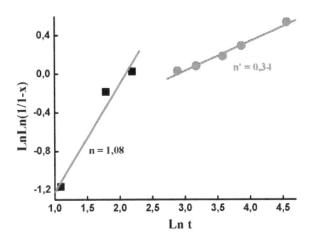

Figure 23. Johnson-Mehl-Avrami plot of the ball-milled $Fe_{57}Co_{21}Nb_7B_{15}$ versus milling time [27, 66].

8.3. FeCoNbB powders

The mixing kinetics of the $Fe_{57}Co_{21}Nb_7B_{15}$ powders can be described by two stages [27, 66] with different Avrami parameters n = 1.08 and n' = 0.34 (Fig. 23). The lower values of the Avrami parameter can be ascribed to the presence of both Nb and B which favour the grain size refinement and the formation of a highly disordered state. For the $(Fe_{50}Co_{50})_{62}Nb_8B_{30}$ mixture, two stages have been obtained with different Avrami parameter values n_1 = 1.41 and n_2 = 0.34 [2]. The former value is comparable to those obtained for the Finemet and Nanoperm [67]. However, it is higher than that obtained during the crystallization of the amorphous FeCoNbB alloy where α-(Fe,Co) nanocrystals with grain size of 15 nm are distributed in the amorphous matrix [65]. Bigot et al. have obtained a value of n = 1.5 for the nanocrystallization of the Finemet [68]. Comparable kinetics parameters have been obtained in the Ni-15Fe-5Mo (n = 1.049 and k = 0.57) [69]. The important fraction of structural defects which is introduced during the milling process favours the phase formation through the diffusion at the surface which is dominant, at lower temperatures, in comparison to the diffusion by the grain boundaries and the lattice parameter (vacancy's diffusion).

9. Conclusion

Thermal analysis is widely used in the reaction study of the mechanically alloyed powder particles because of the obtained metastable disordered structures. Hence, thermal annealing leads to the relaxation of the introduced stresses during the milling process. The heat effects are dependent on the structural and microstructural properties of the ball-milled powders.

Author details

Safia Alleg* and Saida Souilah
Badji Mokhtar Annaba University, Department of Physics, Laboratoire de Magnétisme et Spectroscopie des Solides (LM2S) B.P. 12, 23000 Annaba, Algeria

Joan Joseph Suñol
Dep. De Fisica, Universitat de Girona, Campus Montilivi, 17071 Girona, Spain

Acknowledgement

Prof. Safia Alleg is grateful to the University of Girona-spain for the financial support as invited professor. Financial support from AECID A/016051/08 and AECID A/025066/09 projects is acknowledged. Financial support from WLI Algeria is acknowledged.

10. References

[1] Suryanarayana C (2004) Mechanical alloying and milling. Marcel Dekker. 457-p

[2] Alleg S, Azzaza S, Bensalem R, Suñol JJ, Khene S, Fillion G (2009) Magnetic and structural studies of mechanically alloyed $(Fe_{50}Co_{50})_{62}Nb_8B_{30}$ powder mixtures. J. Alloys Compd. 482: 86-89.

[3] Moumeni H, Alleg S, Djebbari C, Bentayeb FZ, Greneche JM (2004) Synthesis and characterization of nanostructured FeCo alloys. J. Mat. Sci. 39: 5441-5443.

[4] Bensebaa N, Alleg S, Bentayeb FZ, Bessais L, Greneche JM (2005) Microstructural characterization of Fe-Cr-P-C powder mixture prepared by ball milling. J. Alloys Compd. 388:41-48.

[5] Bentayeb FZ, Alleg S, Bouzabata B, Greneche JM (2005) Study of alloying mechanisms of ball milled Fe-Cr and Fe-Cr-Co powders. J. Magn. Magn. Mat. 288: 282-296.

[6] Tebib W, Alleg S, Bensalem R, Greneche JM (2010) Structural study of the mechanically alloyed Fe-P powders. Int. J. Nanoparticles 3:237-244.

[7] Azzaza S, Alleg S, Moumeni H, Nemamcha AR, Rehspringer J L, Greneche J M (2006) Magnetic properties of nanocrystalline ball milled Fe and $Fe_{50}Co_{50}$ alloy. J. Phys.: Condens. Matter 18 : 7257-7272.

[8] Tebib W, Alleg S, Bensebaa N, Bentayeb FZ, Suñol JJ, Greneche JM (2008) Structural characterization of nanostructured Fe-8P powder mixture. J. Nanosci. Nanotechnol. 8:2029-2036.

---

* Corresponding Author

[9] Alleg Safia, Bentayeb Fatima Zohra, Djebbari Chafia, Bessais Lotfi, Greneche Jean Marc (2008) Effect of the milling conditions on the formation of nanostructured Fe-Co powders. Phys. Stat. Sol. (a) 205: 1641-1646.

[10] Alleg S, Ibrir M, Fenineche NE, Azzaza S, Suñol JJ(2010) Magnetic and structural characterization of the mechanically alloyed $Fe_{75}Si_{15}B_{10}$ powders. J. Alloys Compd. 494: 109-115.

[11] Bansal C, Gao ZQ, Hong L B, Fultz B (1994) Phases and phase stabilities of Fe_3X alloys (X=Al, As, Ge, In, Sb, Si, Sn, Zn) prepared by mechanical alloying. J. Appl. Phys. 76:5961–5966.

[12] Macrí PP, Enzo S, Cowlam N, Frattini R, Principi G, Hu WX (1995) Mechanical alloying of immiscible $Cu_{70}TM_{30}$ alloys (TM = Fe,Co). Philosophical Magazine Part B 71:249-259.

[13] Bentayeb FZ, Alleg S, Greneche J M (2007) Structural and microstructural study of Fe-31Cr-12Co mixture prepared by ball milling. J. Alloys Compd. 434: 435-477.

[14] Dekhil L, Alleg S, Suñol JJ, Greneche JM (2009) X-rays diffraction and Mössbauer spectrometry studies of the mechanically alloyed Fe-6P-1.7C powders. Adv. Pow. Technol. 20:593-597.

[15] Azzaza S (2006) Magister. Badji Mokhtar Annaba University, Algeria.

[16] Sherif El-Eskandarany M, Saida J, Inoue A (2002) Amorphization and crystallization behaviours of glassy $Zr_{70}Pd_{30}$ alloys prepared by different techniques. Acta Mater. 50:2725–2736.

[17] Kissinger HE (1957) Reaction kinetics in differential thermal analysis. Anal. Chem. 29:1702–1706.

[18] Greer AL (1982) Crystallization kinetics of $Fe_{80}B_{20}$ glass. Acta Metall. 30:171–192.

[19] Henderson D W (1979) Thermal analysis of non-isothermal crystallization kinetics in glass forming liquids. J. Non-Cryst. Solids 30:301–315.

[20] Galwey AK, Brown ME (1998) Kinetic background to thermal analysis and calorimetry. Handbook of thermal analysis and calorimetry: principles and practice. Brown ME, editor. Elsevier Science B.V. pp. 147-224.

[21] Christian JW (1975) The Theory of Transformations in Metals and Alloys. Pergamon, Oxford p. 542.

[22] Avrami M (1939) Kinetics of phase change I. J. Chem. Phys. 7: 1103-1112.

[23] Avrami M (1940) Kinetics of phase change II. J. Chem. Phys. 8: 212-224.

[24] Avrami M(1941) Kinetics of phase change III. J. Chem. Phys. 9: 177-184.

[25] Kolmogorov AN (1937) Statistical theory of crystallization of metals. Bull. Acad. Sci. USSR, Phys. Sci.1:355-359.

[26] Moumeni H, Alleg S, Greneche JM (2006), Formation of ball-milled Fe-Mo nanostructured powders. J. Alloys Compd. 419: 140-144.

[27] Souilah S, Alleg S, Djebbari C, Suñol JJ (2012) Magnetic and microstructural properties of the mechanically alloyed $Fe_{57}Co_{21}Nb_7B_{15}$ powder mixture. Mat. Chem. Phys. 132: 766-772.

[28] Alleg S, Hamouda A, Azzaza S, Suñol JJ, Greneche JM (2010) Solid state amorphization transformation in the mechanically alloyed $Fe_{27.9}Nb_{2.2}B_{69.9}$ powders. Mat. Chem. Phys. 122: 35-40.

[29] Alleg S, Bensalem R (2011) Nanostructured Fe-based Mixtures Prepared by Mechanical Alloying. In: Jason M. Barker, editor, Powder Engineering, Technology and applications, Nova Science Publishers: pp. 81-124.

[30] Louidi S, Bentayeb FZ, Suñol JJ, Escoda L (2010) Formation study of the ball-milled $Cr_{20}Co_{80}$ alloy. J. Alloys Compd. 493: 110-115.

[31] Loudjani Nadia, Bensebaa Nadia, Alleg Safia, Djebbari Chaffia, Greneche Jean Marc (2011) Microstructure characterization of ball-milled $Ni_{50}Co_{50}$ alloy by Rietveld method. Phys. Status Solidi A 208:2124-2129.

[32] Calka A, Radlinski AP (1986) The effect of surface on the kinetics of crystallization of Pd-Si glassy metals. MRS Proceedings 80:195-201.

[33] Gibson MA, Delamore GW (1987) Crystallization kinetics of some iron-based metallic glasses. J. Mater. Sci. 22:4550-4557.

[34] Cao M G, Fritsch HU, Bergmann HW (1985). Thermochim. Acta 83:23.

[35] Lü L, Lai M (1998) Mechanical alloying. Kluwer Academic Publishers. 273 p.

[36] Ibrir M (2011) PhD Thesis. Badji Mokhtar Annaba University, Algeria.

[37] Moumeni H, Alleg S, Greneche JM (2005) Structural properties of $Fe_{50}Co_{50}$ nanostructured powder prepared by mechanical alloying. J. Alloys Compd. 386: 12-19.

[38] Moumeni Hayet, Nemamcha Abderrafik, Alleg Safia, Greneche Jean-Marc (2010) Stacking faults and structure analysis of ball-milled Fe-50%Co powders. Mat. Chem. Phys. 122:439-443.

[39] Brüning R, Samwer K, Kuhrt C, Schultz L (1992) The mixing of iron and cobalt during mechanical alloying. J. Appl. Phys. 72:2978-2983.

[40] Sorescu M, Grabias A (2002) Structural and magnetic properties of $Fe_{50}Co_{50}$ system. Intermetallics 10:317-321.

[41] Lutterotti L (2000) MAUD CPD Newsletter (IUCr) 24.

[42] Lutterotti L, Matthies S, Wenk HR (1999) MAUD: a friendly Java program for material analysis using diffraction. IUCr: Newsletter of the CPD, 21:14-15.

[43] Li S, Wang K, Sun L and Wang Z (1992) Simple model for the refinement of nanocrystalline grain size during ball milling. Scr. Metall. Mater. 27: 437-442

[44] Börner I, Eckert J (1997) Nanostructure formation and steady-state grain size of ball-milled iron powders. Mat. Sci. Eng. A226-228: 541-545.

[45] Fecht HJ (1994) Nanophase Materials. In:Hadjipanayis G C, Siegel R W, editors. 260:125-132

[46] Zhou GF, Bakker H (1994) Atomically disordered nanocrystalline Co_2Si by high-energy ball milling. J. Phys.: Condens. Matter. 6:4043–4052.

[47] McHenry ME, Gallagher K, Johnson F, Scott JH, Majetich SA (1996) Recent advances in the chemistry and physics of fullerenes and related materials. In: Kadish KM, Ruoff RS, editors. PV96-10, ECS Symposium Proceedings, Pennington, NJ, p. 703.

[48] Host J J, Teng M H, Elliot B R, Hwang J H, Mason T O, Johnson D L (1997) Graphite encapsulated nanocrystals produced using a low carbon:metal ratio. J. Mat. Res. 12: 1268-1273.

[49] Krill C E, Merzoug F, Krauss W and Birringer R (1997) Magnetic properties of nanocrystalline Gd and W/Gd. NanoStruct. Mater 9:455-460.

[50] Massalski T (1990) Binary alloy phase diagrams, Materials Park OH: ASM International.

[51] Turgut Z, Huang MQ, Gallagher K, McHenry ME (1997) Magnetic evidence for structural phase-transformations in Fe-Co alloy nanocrystals produced by a carbon arc. J. Appl. Phys. 81: 4039-4041.

[52] Persiano AIC, Rawlings RD (1991) Effect of niobium additions on the structure and magnetic properties of equiatomic iron cobalt alloys. J. Mat. Sci. 26: 4026-4632.

[53] Bensalem R, Younes A, Alleg S, Souilah S, Azzaza S, Suñol JJ, Greneche JM (2011) Solid state amorphisation of mechanically alloyed Fe-Co-Nb-B alloys. Int. J. Nanoparticles 4: 45-52.

[54] Alleg S, Souilah S, Bensalem R, Younes A, Azzaza S, Suñol JJ(2010) Structural characterization of the mechanically alloyed $Fe_{57}Co_{21}Nb_7B_{15}$ powders. Int. J. Nanoparticles 3: 246-256.

[55] Alleg S, Souilah S, Achour Y, Suñol JJ, Greneche JM (2012) Effect of the Nb content on the amorphization process of the mechanically alloyed Fe-Co-Nb-B powders. J. Alloys Compd. 536S:S394-S397.

[56] Blazquez JS, Conde A, Greneche JM (2002) Mössbauer study of FeCoNbBCu hitperm-alloys. Appl. Phys. Letters 81:1612-1614.

[57] Miglierini M, Greneche JM (1997) Mössbauer spectrometry of Fe(Cu)MB-type nanocrystalline alloys II: the topography of hyperfine interactions in Fe(Cu)ZrB alloys. J. Phys.: Condens. Matter 9:2321-2347.

[58] Suñol JJ, Güell JM, Bonastre J, Alleg S (2009) Structural study of nanocrystalline Fe-Co-Nb-B alloys prepared by mechanical alloying. J. Alloys Compd. 483: 604-607.

[59] Gercsi Zs, Mazaleyrat F, Kane SN, Varga LK (2004) Magnetic and structural study of $(Fe_{1-x}Co_x)_{62}Nb_8B_{30}$ bulk amorphous alloys. Mater. Sci. Eng. A 375–377: 1048-1052.

[60] Gloriant T, Suriñach S, Baró MD (2004) Stability and crystallization of Fe-Co-Nb-B amorphous alloys. J. Non-Crystal. Sol. 333: 320-326.

[61] Hernando A, Navarro I, Gorría P (1995) Iron exchange-field penetration into the amorphous interphases of nanocrystalline materials. Phys. Rev. B 51:3281-3284.

[62] Suzuki K, Cadogan JM, Sahajwalla V, Inoue A, Masumoto T (1996) $Fe_{91}Zr_7B_2$ soft magnetic alloy. J Appl. Phys. 79: 5149-5151.

[63] Liebermann HH, Marti J, Martis RJ, Wong CP (1989) The effect of microstructure on properties and behaviours of annealed $Fe_{78}B_{13}Si_9$ amorphous alloy ribbon. Metall. Trans. A 20:63-70.

[64] Alleg S, Rihia G, Bensalem R, Suñol JJ (2009) Structural evolution of the ball-milled $Ni_{70}P_{30}$ powders. Ann. Chim. Sci. Mat. 34:267-273.

[65] Blazquez JS, Conde CF, Conde A (2001) Crystallization process in $(FeCo)_{78}Nb_6(BCu)_{16}$ alloys. J. Non-Cryst. Solids 287:187-192.

[66] Souilah S (2012) PhD thesis. Badji Mokhtar Annaba University, Algeria.

[67] McHenry ME, Willard MA, Laughlin DE (1999), Amorphous and nanocrystalline materials for applications as soft magnets. Prog. Mat. Sci. 44:291-433.

[68] Bigot J, Lecaude N, Perron JC, Milan C, Ramiarinjaona C, Rialland JF (1994) Influence of annealing conditions on nanocrystallization and magnetic properties in $Fe_{73.5}Cu_1Nb_3Si_{13.5}B_9$ alloy. J. Magn. Magn. Mater. 133: 299-302.

[69] Shen SY, Hng HH, Oh JT (2004). Mater. Letter 58:2824.

Chapter 2

Application of Differential Scanning Calorimetry to the Characterization of Biopolymers

Adriana Gregorova

Additional information is available at the end of the chapter

1. Introduction

Generally, polymers can be classified according to their thermal and mechanical properties into thermoplastics, thermosets and elastomers. Thermoplastics are amorphous or semi-crystalline polymers that soft or melt during heating and solidify during cooling. The heating/cooling/heating process can be repeated without perceptible changes in thermal and mechanical properties of thermoplastics. Thermosets during heating undergo chemical changes and this process is irreversible. Elastomers can be vulcanized (cross-linked under assistance of heat, light, or special chemicals like sulfur, peroxides) that makes them reversibly stretchable for small deformations but vulcanization is the irreversible process.

The resulted properties of polymer materials and mixtures depend on the chemical and physical properties of neat polymers, additives as well as the used processing methodology. Differential scanning calorimetry (DSC) is a physical characterization method used to study thermal behavior of neat polymers, copolymers, polymer blends and composites. Generally, the non-isothermal DSC is used for the identification of neat basic polymers as well as the determination of their purity and stability. Amorphous polymers exhibit a glass transition temperature and semi-crystalline polymers may possess the glass transition temperature, a crystallization temperature, a melting temperature with various crystallization and melting enthalpies. However, these properties alter by both a presence of additives and applied polymer processing methodologies. Basically, a small quantity of sample (up to 10 mg) in pan from various materials (e.g. aluminum pan) and empty pan (reference) are treated under a defined temperature program (various combinations of thermal scans-heating/cooling, and isothermal cycles), a pressure (stable) and an atmosphere (inert or reactive). Principally, sample and reference are maintained at the same temperature, while any transition occurred in the sample needs an energy supply, which is recorded by the DSC as a rate dQ/dt against a temperature or a time. The DSC is the thermal analysis mainly used to determine a first-order transition (melting) and a second order endothermic transition (glass transition). The sudden change in the specific heat value, C_p corresponds with the glass transition temperature as follows (Bower, 2002):

$$\frac{dQ}{dt} = mC_p \qquad (1)$$

where m is the mass of the sample.

However, the determination of the glass transition of polymers with a high crystallinity content is limited. The first-order transitions such as the crystallization of a polymer during a heating (cold crystallization) or a cooling cycle (crystallization) and a melting of polymer crystals can be described by the following formula (Bower, 2002):

$$\frac{dQ}{dt} = \kappa \Delta T = \kappa \dot{T}(t - t_0) + \frac{dQ}{dt}\bigg|_{t_0} \qquad (2)$$

where κ is a thermal conductance between a sample holder and a sample, $\dot{T}$ is a temperature increase rate, and t_0 is the start of transition.

Figure 1 shows the example of thermal transitions occurring in the injection molded sample of poly(lactic acid) (PLA) such as the glass transition, the cold crystallization and the melting. PLA is a thermoplastic aliphatic semi-crystalline biodegradable polyester. The presented molded sample had been cooled very rapidly during the processing (injection molding), so as the consequence during the second heating cycle appeared the cold crystallization peak.

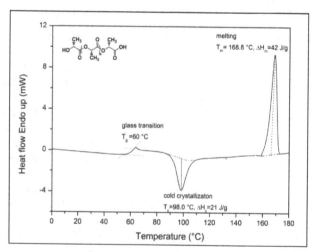

Figure 1. DSC thermogram of commercial poly(lactic acid) with $\overline{Mw}$ = 70 400 and PDI = 1.8 detected during 2nd heating cycle (0-180°C, 10°C/min, N₂ atmosphere)

There are two types of DSC systems: 1) heat-flux (sample and reference pans are in an identical furnace block) and 2) power compensation (sample and reference pans are in two separate furnace blocks). From the practical point of view, it is important to pay attention to issues influencing an accuracy of results as follows:

- an instrument calibration, baseline subtractions,
- a selection of working gas (N_2, He, O_2),
- a selection of pans (e.g. Al-, Pt-, Ni-, Cu-, Quartz-pans, hermetic or non-hermetic pans),
- a proper thermal contact between sample and pans,
- a temperature program (heating cycle usually should start about 50°C under and finish about 10-20°C above the expected measured transition temperature),
- a sufficient slow scanning rate (to avoid the neglecting of the requested thermal transition),
- a sufficient purity and source of sample (neat polymer, polymer blend, composite, before or after processing, kind of the processing).

The aim of this chapter is to show some examples of the practical use of the DSC within the investigation of an amorphous biopolymer – lignin and semi-crystalline biodegradable polymer – poly(lactic acid) as well as to discuss the dependence of the thermal thermal properties on the value of the molecular weight of polymer, the polymer processing methodology and the presence of additives in the polymer mixtures.

2. Effect of molecular weight on glass transition temperature

Amorphous and semicrystalline polymers undergo a phase change from a glassy to rubbery stage at a glass transition temperature (Tg).

At Tg the segmental mobility of molecular chains increases and a polymer is more elastic and flexible. The value of Tg is dependent on the various factors such as a molecular weight of polymer, a presence of moisture, a presence of the crystalline phase (in the case of semicrystalline polymers). The dependence of Tg on a number-average molecular weight is described by Flory-Fox equation:

$$T_g = T_g^\infty + \frac{K}{M_n} \tag{3}$$

where T_g^∞ is a glass transition for polymer with the infinite number-average molecular weight, K is an empirical parameter related to the free volume in polymer and M_n is a number-average molecular weight of polymer.

2.1. Thermal properties of Kraft lignin extracted with organic solvents

In this sub-chapter, an example of the effect of various extraction solvents on molecular weight properties and thermal properties of Kraft lignin is shown.

Lignin is polydisperse amorphous natural polymer consisting of branched network phenylpropane units with phenolic, hydroxyl, methoxyl and carbonyl groups. Its molecular weight properties as well as functional groups depend on its genetic origin and used isolation method. Differential scanning calorimetry is the useful method to determine its glass transition temperature. The value of Tg depends on the molecular weight, the thermal treatment, the humidity content and the presence of various contaminants in lignin sample.

Generally, phenyl groups together with the cross-linking restrict the molecular motion of lignin as an amorphous polymer in contrast to propane chains. Moreover, the intermolecular hydrogen bonding decrease Tg in the contrast to the methoxyl groups (Hatakeyama & Hatakeyama, 2010). Lignin might be defined as a natural polymeric product produced by the enzymatic dehydrogenation polymerization of the primary methoxylated precursors such as p-coumaryl-, coniferyl- and sinapyl- alcohols (Figure 2).

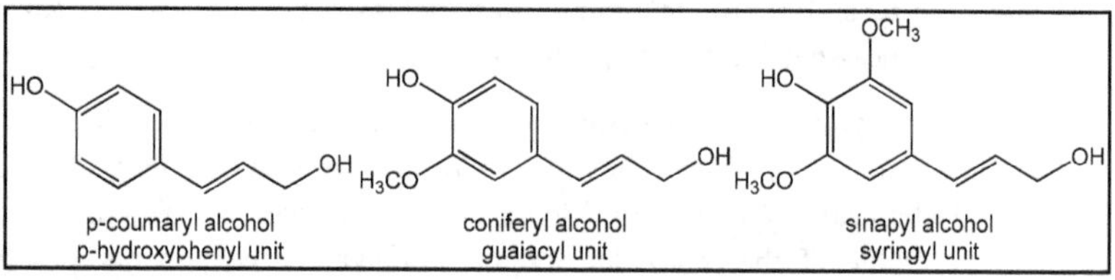

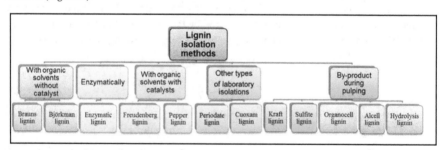

Figure 2. Lignin monomer building units

The structure of lignins depends on their natural origin and also on the external and internal conditions existing during lignin macromolecule synthesis and isolations. The large heterogeneity of lignin's structures makes it difficult to determine the overall structure of lignin. High variability of substituents on phenyl propane unit together with auto-coupling reaction gives rise to different lignin's structures depending on its origin and isolation method (Figure 3).

Figure 3. Lignin isolation methods

Kraft lignin used in this study was isolated from commercial spent pulping black liquor through the acidification with 98% sulphuric acid to pH=2 (Zellstoff Pöls AG, Austria). Precipitated, filtered, washed and dried Kraft lignin was extracted at the room temperature with organic solvents with Hildebrand solubility parameters in the range of 18.5-29.7 MPa$^{1/2}$ (see Table 1) and then again filtered and dried.

Solvent	Chemical formula	Hildebrand solubility parameter (MPa)$^{1/2}$	Polarity index
Dichlormethane	CH$_2$Cl$_2$	20.2	3.1
Tetrahydrofuran	C$_4$H$_8$O	18.5	4.0
Acetone	CH$_3$COCH$_3$	19.7	5.1
1,4-Dioxane	C$_4$H$_8$O$_2$	20.5	4.8
Methanol	CH$_3$OH	29.7	5.1

Table 1. Solvents used for Kraft lignin extraction

The determined thermal and molecular weight properties of Kraft lignins are shown in Table 2. The glass transition temperature (Tg) and the specific heat change (ΔC$_P$) were assessed by the differential scanning calorimetry (DSC) under the nitrogen flow, using the second heating cycle. Molecular weight properties were determined by a gel permeation chromatography (GPC) with the using of tetrahydrofuran as an eluent.

Sample	T$_g$ (°C)	ΔC$_P$ (Jg°C)	$\overline{M_n}$ (g/mol)	$\overline{M_w}$ (g/mol)	PDI
Kraft lignin_acetone	114	0.086	1030	1800	1.7
Kraft lignin_tetrahydrofuran	124	0.222	1170	3150	2.7
Kraft lignin_dichlormethane	59	0.260	750	940	1.3
Kraft lignin_methanol	105	0.368	910	1300	1.4
Kraf lignin_1,4-dioxane	120	0.367	1150	3070	2.7

Table 2. Thermal and molecular weight properties of Kraft lignins extracted in acetone, tetrahydrofuran, dichlormethane, methanol and 1,4-dioxane

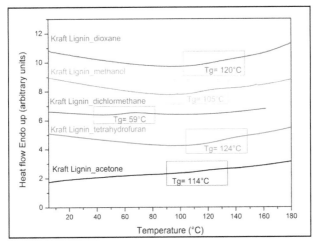

Figure 4. DSC thermograms of Kraft lignin extracted in acetone, tetrahydrofuran, dichlormethane, methanol and 1,4-dioxane detected during second heating scan (5-180°C, 10°C/min, N$_2$ atmosphere)

Figure 4 shows the thermograms of the individual Kraft lignins extracted with various organic solvents.

As can be seen from the results, the extraction as the last step used during the isolation process of Kraft lignin has a big effect on molecular as well as thermal properties of lignin.

2.2. Thermal properties of Poly(lactic acid) synthetized through azeotropic dehydration condensation

This sub-chapter shows the connection between PLA structure, its molecular weight properties and its thermal properties.

Poly(lactic acid) (PLA) is a biodegradable, thermoplastic, aliphatic polyester, which monomer can be derived from annually renewable resources. The glass transition temperature value is an important attribute that influences viscoelastic properties of PLA. The increase of the ambient temperature above Tg of PLA causes the sharp loss of its stiffness. The Tg values of PLA are influenced by its molecular weight, crystallinity, thermal history during processing, character of the side-chain groups and the presence of additives in the composition. The DSC analysis is one of the suitable methods to characterize the effect of the modification of PLA reactive side-chain groups on its thermal properties.

It is worth to mention that the melting temperature and the heat of fusion of polymers are influenced by thermal history applied during the polymer synthesis or processing. Therefore DSC results derived from 1st heating cycle give information concerning an actual state of polymer crystals and the application of cooling cycle erase the previous thermal history, e.g. annealing during processing. Some semi-crystalline polymers with the slow crystallization ability like poly(lactic acid) do not have time to crystallize during cooling and thus crystallize during 2nd heating cycle (cold crystallization) and consequently the melting peak may appear as double peak due to the content of different kinds of crystals. The melting behaviour of PLA is complex with regard to its multiple melting behaviour and polymorphism and has been intensively studied by several authors (Yasiniwa et al., 2004; Yasuniwa et al., 2006; Yasuniwa et al., 2007; Di Lorenzo, 2006).

PLA sample in the following example, marked as *PLA 0*, was synthetized by an azeotropic dehydration condensation in a refluxing boiling m-xylene from 80% L-lactic acid. During the azeotropic dehydration condensation samples *PLA_1-3* were modified by succinic anhydride in the concentration 0.7, 1.3 and 2.5 mol% (Gregorova et al., 2011a). Table 3 summarizes the nomenclature and molecular properties of non-modified PLA and PLA modified with various concentration of succinic anhydride.

Figure 5 shows DSC heating/cooling/heating thermogram of non-modified PLA with the molecular weight of 35 600 g/mol.

Generally, glass transition temperature is determined from the second heating cycle to provide T_g value independent on the thermal history during processing. The modification of PLA side-chain groups by succinic anhydride influenced not just molecular weight properties of PLA but also their thermal properties such as the glass transition temperature (T_g), the melting temperature (T_m) (in this case T_m was determined as the peak temperature of the melting peak) and the crystallinity (see Figure 6. and Table 4). As an adequate indicator of the crystallinity was chosen the specific heat of fusion, calculated as follows:

$$\Sigma \Delta H = (\Delta H_{m1} + \Delta H_{m2}) - \Delta H_c \qquad (4)$$

where ΔH_{m1} and ΔH_{m2} are enthalpy values of the first and second melting peak, ΔH_c is the enthalpy of cold crystallization.

Sample	Concentration of succinic anhydride (mol%)	$\overline{M}_n$ (g/mol)	$\overline{M}_w$ (g/mol)	PDI
PLA_0	0	21400	35600	1.7
PLA_1	0.7	1950	3200	1.6
PLA_2	1.3	5600	9300	1.7
PLA_3	2.5	7000	13000	1.9

Table 3. Description of PLA samples and their molecular properties determined by GPC in chloroform

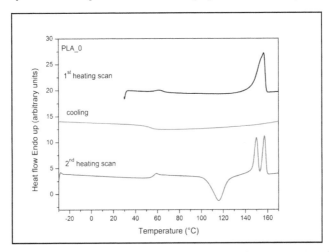

Figure 5. DSC thermogram of PLA_0 detected during heating/cooling/heating scan (30-170°C, 170-0°C, -30-170°C, 10°C/min, N₂ atmosphere)

By the comparison of the content of the crystalline phase determined from 1st heating and 2nd heating cycle, it can be seen that PLA samples during second heating cycle exhibit an amorphous character despite of the initially crystalline character determined from 1st heating scan. A thermal history is very important issue that influence the arrangement of amorphous/crystalline phase and consequently influence the physico-mechanical properties of poly(lactic acid).

| Sample | 1st heating cycle | | | | | 2nd heating cycle | | | | | | | |
	T_{m1} (°C)	ΔH_{m1} (J/g)	T_{m2} (°C)	ΔH_{m2} (J/g)	$\Sigma\Delta H$ (J/g)	T_g (°C)	T_c (°C)	ΔH_c (J/g)	T_{m1} (°C)	ΔH_{m1} (J/g)	T_{m2} (°C)	ΔH_{m2} (J/g)	$\Sigma\Delta H$ (J/g)
PLA_0	157	37.8	-	-	37.8	56	116	27.3	150	14.3	157	15.4	2.4
PLA_1	145	18.8	-	-	18.8	47	106	27.1	132	9.5	143	18.6	1.0
PLA_2	143	10.7	152	15.8	26,5	50	107	23.6	140	7.8	151	20.0	4.2
PLA_3	139	7.4	152	13.5	20,9	50	109	25.9	139	9.9	150	20.1	4.1

Table 4. Thermal properties of PLA synthetized through the azetropic dehydration condensation from 80% L-Lactic acid and modified by succinic anhydride

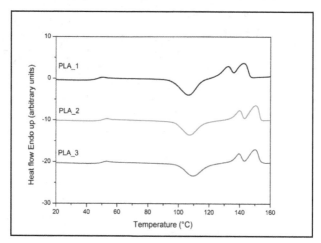

Figure 6. DSC thermograms of PLA samples with modified side chain groups and various molecular properties detected during second heating scan (-30-170°C, 10°C/min, N₂ atmosphere)

3. Effect of thermal treatment on thermal behavior of poly(lactic acid)

As was already discussed in the previous sub-chapter, PLA is the semi-crystalline polymer with the slow crystallization ability. Mechanical properties as well as gas barrier properties of PLA depend also on its gained crystallinity value. The resulting crystallinity of PLA can be modified by a thermal treatment (annealing) for some time at the crystallization temperature during the thermal processing of a sample. The change of a crystals size and a form during the annealing can be revealed by a X-Ray analysis but the change in the percentage of crystalline phase is detectable also by the DSC analysis. This section describes the progress of the PLA crystalline phase due to the applied annealing treatment. Moreover, the obtained DSC data are supported by a light microscopy study.

The followed data were obtained by the analysis of the thermal compression molded poly(lactic acid) synthetized by the azeotropic dehydration condenstation (PLA_3) (Figure 7).

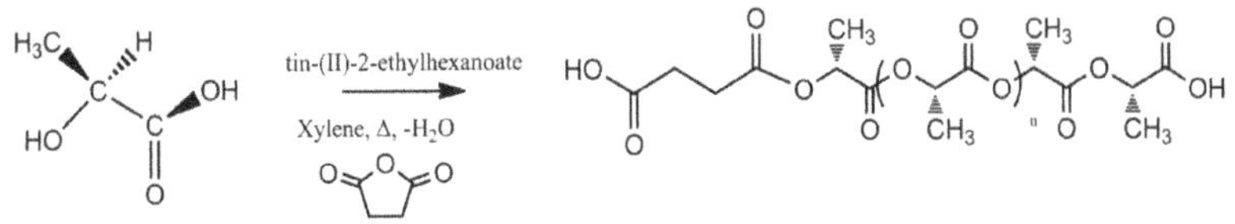

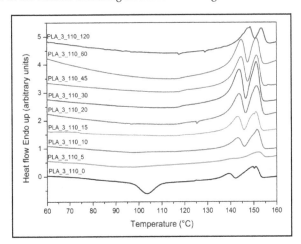

Figure 7. Structure of PLA_3 (PLA synthesized by the azeotropic dehydration condensation and modified by 2.5 mol% succinic anhydride)

The crystallinity value of PLA was modified during thermoprocessing by the thermal annealing at 110°C for 0, 5, 10, 15, 20, 30, 45, 60 and 120 min, respectively and afterwards cooled down to the room temperature. The samples are designated as PLA_3_110_X, where X indicates annealing time.

The clear effect of the thermal annealing on the PLA melting behavior is shown in Figure 8.

Figure 8. DSC thermograms of PLA_3 annealed at 110°C for 0-120 min (1st heating, 30-160°C, 10°C/min, N₂ atmosphere)

The change of the annealing time influenced the value of the specific melting enthalpy ($\Sigma\Delta H$) due to the enabling of a growth of crystals (Table 5).

The crystals morphology of PLA samples annealed at 110°C and various times were investigated by using of the light microscope with crossed polarizers (Figure 9). It can be seen that a shape and dimensions of the created crystals depend on the annealing time.

The DSC as well as the light microscopy analyses showed that the thermo-processed films without the annealing processing step have an amorphous character (Figure 9a), and on other side the application of the annealing processing step at 110°C during thermoforming instead of a quick direct cooling step (to the room temperature) promotes the growth of crystals. A kind, a size, a thickness, and a content of arisen crystals depend on the annealing temperature and time. DSC data displayed in Table 5 showed that the value of the specific

Sample	1st heating cycle								
	T_{c1} (°C)	ΔH_{c1} (J/g)	T_{m1} (°C)	ΔH_{m1} (J/g)	T_{m1}Peak height (mW)	T_{m2} (°C)	ΔH_{m2} (J/g)	T_{m2}Peak height (mW)	$\Sigma\Delta H$ (J/g)
PLA_3_110_0	104	21.8	139	4.6	0.14	151	20.5	0.39	3.3
PLA_3_110_5	-	-	-	-	-	152	12.6	1.18	12.6
PLA_3_110_10	-	-	143	6.1	0.26	151	15.4	0.67	21.5
PLA_3_110_15	-	-	143	20.1	0.51	151	13.5	0.65	33.6
PLA_3_110_20	-	-	143	19.9	0.76	151	15.9	1.0	35.8
PLA_3_110_30	-	-	144	20.8	1.1	151	15.6	1.4	36.4
PLA_3_110_45	-	-	144	14.5	0.87	151	13.0	1.03	27.5
PLA_3_110_60	-	-	144	15.2	0.87	151	11.8	0.88	27.0
PLA_3_110_120	-	-	149	10.3	0.44	154	5.6	0.44	15.9

Table 5. Thermal properties of PLA_3 films, annealed at 110°C for 0-120 min

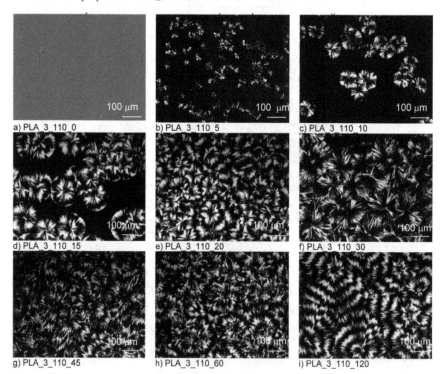

Figure 9. Polarized optical micrographs (magnification 400×) of crystals of polylactic acid modified with succinic anhydride (PLA_3) grown from the melt and annealed at 110°C for 5-120 min

heat of fusion markedly increased up to 15 min of the annealing time, but the extension of the annealing time up to 30 min increased $\Sigma\Delta H$ just slightly and further extension of the annealing time even decreased it. However, light micrographs of *PLA_3* (see Figure 9 b-i) show clear differences of the character of crystals, arisen from the samples annealed under and above 30 min. The application of the longer annealing time caused the creation of overgrowth crystals. The difference in the character of crystals can be also detected by the change of the height of the melting peak and by their shift to the higher temperatures. The value of $\Sigma\Delta H$ of PLA annealed for 120 min (*PLA_3_110_120*) is comparable to that of annealed just for 10 min, however the crystal morphology is markedly different. Furthermore, the change of the crystal morphology was indicated by the increase of the melting temperature (T_{m1} and T_{m2}) about 10 and 3°C, respectively. Also the optical micrograph displayed in Figure 9i showed the difference in the crystal morphology in a comparison to the previous samples annealed at the lower time. As a remark can be highlighted that the crystal morphology has an essential influence on resulting physico-mechanical properties of PLA materials.

4. Thermal stability of biopolymers determined by DSC

4.1. Effect of functional end groups on poly(lactic acid) stability

The intramolecular transesterification with the formation of cyclic oligomers and by-products like acrylic acid, carbon oxide and acetaldehyde is considered as one of the main mechanisms of the PLA thermal degradation. Above 200°C five reaction pathways have been found: intra-and intermolecular ester exchange, cis-elimination, radical and concerted nonradical reactions, radical reactions and Sn-catalyzed depolymerisation (Kopinke et al., 1996). It has been suggested that CH groups of the main chain and the character of functional end groups affect thermal and hydrolytic sensitivity of PLA (Lee et al., 2001; Ramkumar & Bhattacharya, 1998). In our previous work it was shown that thermal sensitivity of PLA might be improved by the modification of its functional end groups (Gregorova et al., 2011a). This sub-chapter shows that the DSC analysis can be used to determine the thermal stability of poly(lactic acid).

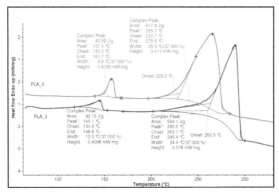

Figure 10. DSC curves of low molecular weight PLA synthetized by azeotropic dehydration condensation (PLA_0) and modified by 2.5 mol.% succinic anhydride (PLA_3), detected by 1st heating cycle from 30 to 350°C at heating rate of 10°C/min, in nitrogen flow.

The obtained DSC data, displayed in Figure 10, showed that the modification of low molecular weight PLA with succinic anhydride caused the decrease of its melting temperature and crystallinity. Furthermore, the detected values of the onset degradation temperature, the degradation temperature in peak and the enthalpy of degradation indicate the improvement of thermal stability, caused by the modification of hydroxyl functional end group by succinic anhydride.

4.2. Stabilizing effect of lignin used as filler for natural rubber

Natural rubber (NR) is highly unsaturated polymer exhibiting poor resistance to oxidation. For the inhibition of the degradation process during thermo-oxidation can be used stabilizers such as phenol and amine derived additives. NR for the production of vulcanized products is mixed with the number of the other compounding ingredients to obtain the desired properties of vulcanizates (e.g sulfur, accelerators, and filler). Lignin is biopolymer that can be used as an active filler for rubber. It was found that some lignins can play dual role in rubber compounds, influencing their mechanical properties as well as their stability [11].

The obtained data were obtained by using of vulcanizates based on natural rubber (NR) and filled with 0, 10, 20 and 30 phr of Björkman beech lignin (Mw= 2000, PDI= 1.2) (Kosikova et al., 2007). Samples are designated as NR_Lignin_X, where X presents concentration of lignin in phr (parts per hundred rubber).

Table 6 shows values of degradation temperature determined as the onset and the peak temperature in dependence on the lignin concentration in natural rubber vulcanizates. It can be seen that lignin used as filler exhibit also the stabilizing effect, while the best stabilizing effect was reached in the case of 20 phr presence of Björkman beech lignin.

Sample	T_{onset} (°C)	T_{peak} (°C)	ΔH (J/g)
NR_Lignin_0	184	326	886
NR_Lignin_10	183	349	833
NR_Lignin_20	301	368	363
NR_Lignin_30	296	364	318

Table 6. DSC data evaluated from 1st heating cycle analysis (30-500°C, 10°C/min, air atmosphere) of vulcanizates based on natural rubber (NR) and NR filled with Björkman beech lignin (Kosikova et al., 2007)

4.3. Stabilizing effect of lignin used as additive in polypropylene

It was already reported that the lignin in the certain circumstances can support the biodegradability of polymer samples (Kosikova et al., 1993a; Kosikova et al., 1993b; Mikulasova&Kosikova, 1999). On the other side lignin with the important functional groups and the low molecular weight with the narrow polydispersity can be used as the stabilizer for polypropylene (Gregorova et al., 2005a). This section shows that DSC is the sensitive method able to determine the stabilizing effect of lignin in polypropylene.

The polypropylene samples, stabilized with Björkman beech lignin (Mw= 2000, PDI= 1.2), used in this example were thermal processed with the injection molding (Gregorova et al., 2005a). Figure 11 shows the change of the onset oxidation temperature (T_{onset}) recorded for polypropylene stabilized with lignin. Generally, additives should be compatible with polymer matrix to keep physico-mechanical properties on the desired level; therefore it is necessary to know the lowest active concentration of the additive. It can be seen that the studied Björkman beech lignin increased T_{onset} about 15-30°C depending on the used concentration. On the base of the obtained mechanical properties of polypropylene/lignin composites, 2 wt% of Björkman beech lignin was determined as the optimal concentration to stabilize polypropylene. It was shown that the higher concentration of non-modified lignin deteriorated the mechanical properties of polypropylene (Gregorova et al., 2005a, Gregorova et al., 2005b).

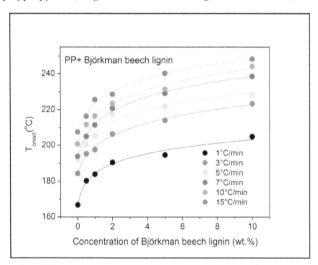

Figure 11. Thermal stability of polypropylene expressed as onset degradation temperature (T_{onset}) in dependence on lignin concentration, heating scan 30 to 500°C, heating rate of 1, 3, 5, 7, 10 and 15 °C/min, oxygen flow (Gregorova et al., 2005a).

5. Thermal properties of poly(lactic acid) composites

The incorporation of filler in PLA may change its crystallization behaviour and consequently its thermal properties. Some filler, such as wood flour or wood fibers, promote the transcrystallization and thus modify crystalline morphology of PLA (Mathew et al.; 2005 Pilla et al., 2008; Matthew et al., 2006; Hrabalova et al. 2010). This section describes the ability of hydrothermally pretreated beech flour to support a nucleation of PLA. Moreover, the effect of quick cooling and thermal annealing during thermal processing of PLA films is recorded.

The sample used in this section were thermoplastic processed compounds of commercial poly(lactic acid) (PLA 7000D, NatureWorks LLC, USA) plasticized with 10 vol% of polyethylene glycol 1500 and filled with 30 wt% of hydrothermally pretreated beech flour (Gregorova et al., 2011b). Composite films were prepared by thermal molding in press at 160°C, 10 MPa for 5 min and by modification of cooling process were prepared two morphologies: amorphous (quick cooling) and semi-crystalline (thermal annealing at 100°C for 45 min). The samples are designated as pPLA_X_100_Y, where X indicates filler (0-no filler, WF- hydrothermally pretreated beech flour) and Y shows annealing time.

The thermal behavior of quenched and annealed PLA composites, investigated by differential scanning calorimetry (heating cycle from 20 to 180°C, 10°C/min, 60 ml/L nitrogen flow) is summarized in Table 11 and shows that both filler incorporation of wood flour and thermal annealing influenced melting behavior and crystallinity of PLA composites. Specific melting enthalpy as an indicator of crystallinity degree of PLA in the composite was calculated as follows:

$$\Sigma \Delta H = \frac{(\Delta H_{m1} + \Delta H_{m2}) - \Delta H_c}{\upsilon} \tag{5}$$

where ΔH_{m1} and ΔH_{m2} are enthalpy values of the first and second melting peak, ΔH_c is the enthalpy of the cold crystallization and v is volume fraction of PLA in the composite.

Sample	1st heating cycle							2nd heating cycle	
	ΔH_c (J/g)	T_c (°)	ΔH_{m1} (J/g)	T_{m1} (°)	ΔH_{m2} (J/g)	T_{m2} (°)	$\Sigma \Delta H$ (J/g)	$\Delta H_{m1}/ \Delta H_{m2}$	T_g (°C)
pPLA_0_100_0	18.5	82	0.4	134	23.7	150	5.6	0.02	35
pPLA_0_100_45	-	-	11.8	142	14.2	152	26.0	0.83	35
pPLA_WF_100_0	14.2	95	1.7	140	12.9	151	0.4	0.13	43
pPLA_WF_100_45	-	-	6.1	148	8.4	155	14.4	0.73	45

Table 7. DSC thermal data of non-annealed and annealed PLA composites determined (Gregorova et al.; 2011b)

Samples that were after melting quickly cooled down to room temperature (quenched) exhibit cold crystallization and the double melting behavior that may be attributed to the melting of the original crystals and those of formed through the cold crystallization from the glassy state (Ling & Spruiell, 2006). The known slow crystallization ability of PLA and quick cooling process caused that quenched samples remained mostly amorphous that was proved by low value of specific enthalpy $\Sigma \Delta H$. Thermograms of annealed samples displayed a marked double melting peak showing high degree of crystallinity (Gregorova et al.;

2011b). The presence of filler marginally decreased specific enthalpy values of PLA. The presence of multiple melting peaks in thermograms of annealed samples can be explained by applied annealing that induce other crystal population, namely α' (initially created with grain like morphology) and α (during annealing created with spherulitic morphology) crystals (Zhang et al., 2008; Pan et al., 2008). Melting temperature for unannealed neat or filled PLA samples were recorded between 134-140°C for the first melting peak and 150-151°C for the second melting peak. The growth of crystals during annealing increased the values of temperature of both melting peaks depending on the mixture composition. The change in the value of the ratio of the first and the second melting peaks indicates the modification of size of the present crystals. The Tg value after an annealing treatment can be taken as an indicator for the occurred changes in an amorphous/crystal ratio but also in PLA/filler interaction level. The increase of an interfacial compatibility between wood filler and poly(lactic acid) can be detected by an shift of a glass transition to the higher temperature (Gregorova & Wimmer, 2012).

6. Conclusions

Differential scanning calorimetry is the method to characterize thermal behavior of polymeric materials on the base of the differences obtained in the heat flow between a sample and a reference under various temperature programs. In the addition to the quality and compositional analyses of polymers, DSC is applicable to the investigation of the thermal changes occurring in polymer systems during chemical reactions (e.g. polymerisation), oxidative degradation, vaporization, sublimation and desorptionThe selection of a proper temperature program is an important issue for the proper DSC analysis (e.g. a position and a shape of melting peak depend inherently on the nature of polymer and on the used heating scan rate). Thermal properties of biopolymers depend on many factors such as their natural origin, purity, composition, processing, thermal treatment, mechanical stressing, and aging. In this chapter, non-isothermal DSC was introduced as an method to investigate thermal properties of biopolymers, namely amorphous lignin and semi-crystalline poly(lactic acid). It can be concluded that DSC is one of the available methods to determine thermal properties of lignin with various molecular weight properties and composition.. Moreover, DSC can serve as a method to determine stabilizing effect of lignin used as an additive in polymer samples. Furthermore, DSC can be used as the quick method to measure melting behavior and the crystallinity of poly(lactic acid). The thermal history during polymer processing as well as the incorporation of some filler (e.g. wood flour) or additives can modify the crystallinity of PLA. The percentage of the crystallinity is one of the most important characteristics that influence its physico-mechanical behavior (stiffness, toughness, brittleness, barrier resistance, thermal stability and optical clarity). DSC is the valuable method for the investigation of thermal properties of biopolymers. However, it is necessary to use also the other additional physical and chemical testing methods to obtain complex data describing biopolymers, such as lignin and poly(lactic acid).

Author details

Adriana Gregorova
Graz University of Technology, Institute for Chemistry and Technology of Materials, Austria

7. References

Bower, B.I. (2002). *An Introduction to Polymer Physics*. Cambridge University Press, New York

Di Lorenzo, M.L. (2006). The Crystallization and Melting Processes of Poly(L-lactic acid). *Macromol. Symp.*, Vol.234, pp. 176-183

Gregorova, A.; Cibulkova, Z.; Kosikova, B.& Simon P. (2005a). Stabilization effect of lignin in polypropylene and recycled polypropylene. *Polymer Degradation and Stability*, Vol. 89, No. 3, pp. 553-558, ISSN 0141-3910

Gregorova, A.; Kosikova, B.& Osvald, A. (2005b). The study of lignin influence on properties of polypropylene composites. *Wood Research*, Vol. 50, No. 2, pp. 41-48, ISSN 1336-4561

Gregorova A.; Schalli M.& Stelzer F. (2011a). Functionalization of polylactic acid through azeotropic dehydrative condensation. *19th Annual Meeting of the BioEnvironmental Polymer Society BEPS, Book of Abstracts, PO-4*, ISBN 978-3-9502992-3-6, Vienna Austria, September 2011

Gregorova, A.; Sedlarik, V.; Pastorek, M.; Jachandra, H.& Stelzer, F. (2011b). Effect of compatibilizing agent on the properties of highly crystalline composites based on poly(lactic acid) and wood flour and/or mica. *Journal of Polymers and the Environment*, Vol. 19, No.2, pp. 372-381, ISSN 1566-2543

Gregorova, A& Wimmer R. (2012). Filler-Matrix Compatibility of Poly(lactic acid) Based Composites. In: Piemonte V., Editor. *Polylactic Acid: Synthesis, Properties and Applications*, Piemonte, V., pp. 97-119, Nova Science Publishers NY, ISBN 978-1-62100-348-9

Hatakeyama, H. & Hatakeyama T. (2010). Lignin Structure, Properties and Applications. In: *Biopolymers Lignin, Proteins, Bioactive Nanocomposites*, Abe A., Dusek K., Kobayashi S., pp. 11-12, Springer-Verlag Berlin Heidelberg

Hrabalova, M.; Gregorova, A.; Wimmer, R.; Sedlarik, V.; Machovsky, M.& Mundigler N. (2010). Effect of Wood Flour Loading and Thermal Annealing on Viscoelastic Properties of Poly(lactic acid) Composite Films. *Journal of Applied Polymer Science*, Vol. 118, No. 3, pp. 1534-1540, ISSN 1097-4628

Kopinke, F.D.; Remmler, M.; Mackenzie, K.; Möder, M.& Wachsen, O. (1996). Thermal decomposition of biodegradable polyesters-II. Poly(lactic acid). *Polymer Degradation and Stabilility*, Vol. 53, No. 3, pp. 329-342, ISSN 0141-3910

Kosikova, B.; Kacurakova, M.& Demianova V. (1993a). Photooxidation of the composite lignin/polypropylene films. *Chemical Papers*, Vol.47, pp. 132-136, ISSN 0366-6352

Kosikova, B.; Demianova, V.& Kacurakova, M. (1993b). Sulphur-free lignins as composites of polypropylene films. *Journal of Applied Polymer Science*, Vol. 47, No. 6, pp. 1065-1073, ISSN 1097-4628

Kosikova, B.; Gregorova, A.; Osvald, A.& Krajcovicova, J. (2007). Role of Lignin Filler in Stabilization of Natural Rubber-Based Composites. *Journal of Applied Polymer Science*, Vol. 103, No. 2, pp. 1226-1231, ISSN 1097-4628

Lee, S-H.; Kim, S.H.; Han, Y-K.& Kim Y.H. (2001). Synthesis and degradation of end-group-functionalized polylactide. *Journal of Polymer Science Part A: Polymer Chemistry*, Vol. 39, No. 7, pp. 973-985, ISSN 1099-0518

Ling, X.& Spruiell, J.E. (2006). Analysis of the complex thermal behaviour of poly(L-lactic acid) film. I. Samples crystallized from the glassy state. *Journal of Polymer Science Part B: Polymer Physics*, Vol. 44, No. 22, pp. 3200-3214, ISSN 1099-0488

Mathew, A.P.; Oksman, K.& Sain, M. (2005). Mechanical properties of biodegradable composites from poly lactic acid (PLA) and microcrystalline cellulose (MCC). *Journal of Applied Polymer Science*, Vol.97, No. 5, pp. 2014-2015, ISSN 1097-4628

Mathew, A.P.; Oksman, K.& Sain, M. (2006). The effect of morphology and chemical characteristics of cellulose reinforcements on the crystallinity of polylactic acid. *Journal of Applied Polymer Science*, Vol. 101, No. 1, pp. 300-310, ISSN 1097-4628

Mikulasova, M.& Kosikova, B. (1999). Biodegradability of lignin-polypropylene composite films. *Folia Microbiologica*, Vol. 44, pp. 669-672, ISSN 0015-5632

Pan, P.; Zhu, B.; Kai, W.; Dong, T.& Inoue, Y. (2008). Polymorphic Transition in Disordered Poly(L-lactide) Crystals Induced by Annealing at Elevated Temperatures. *Macromolecules*, Vol.41, No. 12, pp. 4296-4304, ISSN 1520-5835

Pilla, S.; Gong, S.; O'Neil, E.; Rowell, M.& Krzysik, A.M. (2008). Polylactide-pine wood flour composites. *Polymer Engineering Science*, Vol.48, No. 3, pp. 578-587, ISSN 1548-2634

Ramkumar, D.H.S.& Bhattacharya, M. (1998). Steady shear and dynamic properties of biodegradable polyesters. *Polymer Engineering Science*, Vol. 38, No. 9, pp. 1426-1435, ISSN 1548-2634

Yasuniwa, M.; Tsubakihara, S.; Sugimoto, Y.& Nakafuku, C. (2004). Thermal analysis of the double-melting behavior of poly(L-Lactic acid). *Journal of Polymer Science Part B: Polymer Physics*, Vol.42, No. 1, pp. 25-32, ISSN 1099-0488

Yasuniwa, M.; Tsubakihara, S.; Iura, K.; Ono, Y.; Dan, Y.& Takashashi K. (2006). Crystallization behavior of Poly(L-lactic acid). *Polymer*, Vol. 47, No. 21, pp. 7554-7563, ISSN 0032-3861

Yasuniwa, M.; Iura, K.& Dan Y. (2007). Melting behavior of poly(L-lactic acid): Effects of crystallization temperature and time. *Polymer*, Vol. 48, No. 18, pp. 5398-5407, ISSN 0032-3861

Zhang, J.; Tashiro, K.; Tsuji, H.& Domb, A.J. (2008). Disorder-to-Order Phase Transition and Multiple Melting Behavior of Poly(L-lactide) Investigated by Simultaneous Measurements of WAXD and DSC. *Macromolecules*, Vol.41, No. 4, pp. 1352-1357, ISSN 1520-5835

Studies on Growth, Crystal Structure and Characterization of Novel Organic Nicotinium Trifluoroacetate Single Crystals

P.V. Dhanaraj and N.P. Rajesh

Additional information is available at the end of the chapter

1. Introduction

Single crystal growth has a prominent role in the present era of rapid scientific and technical advancement, whereas the application of crystals has unbounded limits. New materials are the lifeblood of solid state research and device technology. Nonlinear optical (NLO) crystals have come upon the materials science scene and are being studied by many research groups around the world. These materials operate on light in a way very analogous to the way of semiconductors which operate on electrons to produce very fast electronic switching and computing circuits.

Organic crystals have compounds with carbon atoms as their essential structural elements. The design and synthesis of organic molecules exhibiting NLO properties have been motivated by the tremendous potential for their applications in the fast developing domains of optoelectronics and photonic technologies. The relevance of the organic materials in the present context is, the delocalized electronic structure of π–conjugated organic compound offers a number of tantalizing opportunities in the applications as NLO materials. Extensive research in the last decades has shown that organic crystals often possess a higher degree of optical nonlinearity than their inorganic counterparts [1, 2]. Some of the advantages of organic materials include inherently high nonlinearity, high electronic susceptibility through high molecular polarizability, fast response time, the ease of varied synthesis, scope for altering the properties by functional substitutions, high damage resistance, relative ease of device processing, etc. Organic materials have another advantage over inorganic materials, in that the properties of organic materials can be optimized by modifying the molecular structure using molecular engineering and chemical synthesis [3]. A very large operating bandwidth modulation in organic electro-optic devices can be obtained through its low dielectric constant at low frequencies. Hence they are projected as forefront candidates for fundamental and applied investigations.

In organic materials, there is a strong correlation between structure and nonlinear properties. Thus, in the case of second order nonlinear effects, it has been established that the macroscopic susceptibility of the materials $\chi^{(2)}$, is related to both the magnitude of the molecular nonlinearities, i.e. the first hyperpolarizability β, and the relative orientation of the molecules in the medium. Therefore, a fundamental limitation for second-order effects to be observed is the non-centrosymmetry requirement, both at the microscopic molecular level and at the macroscopic bulk level. On the other hand, the third order effects described by $\chi^{(3)}$ can be present in any medium. The $\chi^{(3)}$ coefficients are thus essential in centrosymmetric compounds where the second order coefficients equal zero. They are also important in the non-centrosymmetric molecules. Moreover, these $\chi^{(3)}$ coefficients play a part in some experimental determination of $\chi^{(2)}$ coefficients. It is for example the case in electric field induced second harmonic generation [4-6] (EFISHG) experiment or in hyper-Rayleigh technique [7-10] where the two-photon absorption (TPA), which is third order effect, can induce fluorescence thus making imprecise the determination of the β coefficient. An ultimate goal for designing the molecules with large third order nonlinearities is to incorporate them into devices used in all optical signals processing [11, 12]. Nonlinear optical absorption (NOA) has shown its potential application in optical information storage, all optical logic gates, laser radiation protection, and locked laser mode. Interest in searching for NOA materials has been gradually increased. Organic molecules have been the subjects of great attention due to their potential applications in nonlinear optics, optical switching, and light emitting diodes. Indeed, the potential use of organic device materials in optoelectronics is now a very serious matter.

In order to achieve good macroscopic nonlinear response in organic crystals, one requires an increase in the number of π electrons and π delocalization length, so as to lead to high molecular hyperpolarizability and also proper orientation of the molecule in the solid-state structure to facilitate high-frequency conversion efficiency. Effective materials generally contain donor and acceptor groups positioned at either end of suitable conjugation path. The increased effective conjugation and the large π delocalization length have been recognized as the factors leading to the large third order nonlinearities. While the engineering for enhancing second order NLO efficiency is relatively well understood, the need for efficient third order molecules and materials still exists. The design of organic polar crystals for the quadratic NLO applications is supported by the observation that organic molecules containing π electron systems asymmetrized by electron donor and acceptor groups are highly polarizable entities [13]. Donor/acceptor benzene derivatives are sure to produce high molecular nonlinearity. So far, many organic Donor–π–Acceptor (D–π–A) type compounds have been studied theoretically and also experimentally [14]. The studies indicate that the organic D–π–A compounds are highly promising candidates for NLO applications.

Nicotinic acid, a B vitamin also known as niacin, and its derivatives have been studied extensively over the last decade due to their biological and chemical importance. Niacin forms coordination complexes with tin (Sn), which have been found to have better antitumor activity than the well-known cis-platin or doxorubicin [15]. Many of its pharmacological properties are detailed in literature [16-18]. The reported structures of complexes reveal that nicotinic acid and its derivatives acting as bridging ligands through the carboxylate group and pyridyl N atom [19]. We have synthesized the crystalline salt of nicotinium trifluoroacetate and their crystals in monoclinic system were grown by using solution growth technique for the first time. The crystal structure of nicotinium trifluoroacetate in triclinic system has reported by S. Athimoolam and S. Natarajan [20]. Here we report monoclinic polymorph of nicotinium trifluoroacetate, its asymmetric unit contains a protonated nicotinium cation and a trifluroacetate anion. This chapter discusses synthesis, solubility, crystal growth, structural, dielectric and mechanical properties of nicotinium trifluoroacetate (NTF). Thermal properties of NTF were analyzed and compared with that of two nicotinium derivative crystals nicotinium oxalate and nicotinium nitrate monohydrate.

2. Experimental studies

2.1. Synthesis of NTF

NTF was synthesized by the reaction between nicotinic acid (SRL, India) and trifluoroacetic acid (Merck) taken in equimolar ratio. The growth solution was prepared by adding calculated amount of trifluoroacetic acid slowly in saturated aqueous solution of equimolar nicotinic acid at 50 °C. The continuous stirring of the solution for 6 h at constant temperature using a temperature controlled magnetic stirrer yielded the precipitate of crystalline substance of NTF. Repeated crystallization and filtration processes were applied for the purification of the synthesized compound.

2.2. Determination of solubility and metastable zone width

The nucleation studies were carried out in a constant temperature bath (CTB) with cooling facility of accuracy of ±0.01 °C. The solubility at 30 °C was determined by dissolving the recrystallized salt of NTF in 100 ml Millipore water of resistivity 18.2 MΩcm taken in an air tight container. The solution was stirred continuously for 6 h to achieve stabilization using an immersible magnetic stirrer. After attaining the saturation, the concentration of the solute was estimated gravimetrically. The same procedure is repeated for different temperatures (35, 40, 45 and 50 °C).

Metastable zone width is an essential parameter for the growth of large size crystals from solution, since it is the direct measure of the stability of the solution in its supersaturated region. The metastable zone width was measured by adopting the conventional polythermal method [21]. The saturated solution (100 ml) at 30 °C was prepared according to the presently determined solubility data. After attaining the saturation, the solution was filtered by the filtration pump and Whatman filter paper of pore size 11 μm. The solution was preheated to 5 °C above the saturated temperature for homogenization and left at the superheated temperature for about 1 h before cooling. Then it was slowly cooled at a desired cooling rate of 4 °C/h, until the first crystal appeared. The temperature was instantly recorded. The difference between the saturation temperature and nucleation temperature gives the metastable zone width of the system. Then experiment was repeated for different saturation temperatures 35, 40, 45 and 50 °C and the corresponding metastable zone widths were measured. Several runs (3–5 times) were carried out under controlled conditions for the confirmation of the saturation and nucleation points. The measured values of solubility and metastable zone width of NTF are shown in Figure 1. It shows that NTF has good solubility in water and it increases almost linearly with temperature. Hence solution growth could be a better method to grow good quality single crystals of NTF. The value of the metastable zone width depends not only on the temperature but also on the type of the crystal and its physicochemical properties [22]. One can observe that the metastable zone width decreases with increasing temperature.

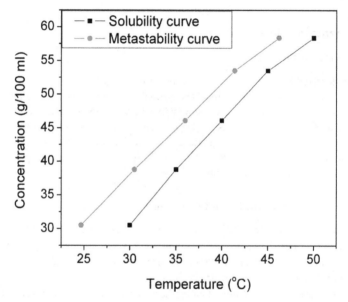

Figure 1. Solubility and metastablity curves of NTF

2.3. Crystal growth

Slow evaporation method was employed for growing the single crystals of NTF. The recrystallized salt was used for the preparation of saturated solution at room temperature (35 °C). The solution was filtered by filtration pump and Whatman filter paper of pore size 11 μm. Then the filtered solution was transferred to a petridish with a perforated lid in order to control the evaporation rate and kept undisturbed in a dust free environment. The single crystal of NTF of size 27 x 12 x 7 mm³ was harvested from mother solution after a growth period of 22 days. The grown single crystal of NTF is shown in the Figure 2.

Morphology of the grown crystals was identified by the single crystal X-ray diffraction studies (Bruker Kappa APEXII). It establishes that the crystal has eight developed faces, out of which $(\overline{1}12)$ and $(1\overline{1}2)$ planes are more prominent. For each face, its parallel Friedal plane is also present in the grown crystal and shown diagrammatically in Figure 3.

Figure 2. Photograph of as-grown NTF crystal

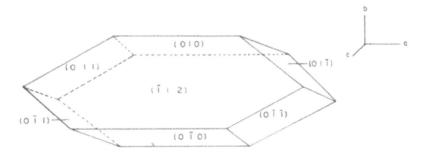

Figure 3. Morphology representation of NTF crystal

3. Analysis of physicochemical studies

3.1. X-ray diffraction analysis

The unit cell parameters and crystal structure of NTF were determined from single crystal X-ray diffraction data obtained with a four-circle Nonius CAD4 MACH3 diffractometer (graphite monochromated, MoKα = 0.71073 Å) at room temperature (293 K). The data reduction was done by using XCAD4 [23] and absorption correction was done by the method of ψ–scan [24]. The structure solution and refinement were performed using SHELXTL 6.10 [25]. The crystal structure of NTF was solved by direct methods, and full-matrix least-squares refinements were performed on F^2 using all the unique reflections. All the non-hydrogen atoms were refined with anisotropic atomic displacement parameters, and hydrogen atoms were refined with isotropic displacement factors. The crystallographic data and structure refinement parameters of NTF crystal are presented in Table 1. The crystal structure and packing diagram of NTF are shown in Figures 4 and 5 respectively. The H atom participating in the N–H bond was located from the difference Fourier map and all other H atoms were positioned geometrically and refined using a riding model with C–H = 0.93 Å and O–H = 0.82 Å with Uiso(H) = 1.2 – 1.5 Ueq (parent atom). The absolute configuration is assigned from the starting materials taken for reaction.

Empirical formula	$C_8H_6F_3NO_4$
Formula weight	237.14
Temperature	293(2) K
Wavelength	0.71073 Å
Crystal system, space group	Monoclinic, $I2/a$
Unit cell dimensions	a = 15.616(5) Å, α = 90° b = 7.455(5) Å, β = 95.74° c = 16.503(5) Å, γ = 90°
Volume	1911.6 Å^3
Z, Calculated density	8, 1.648 g/cm^3
Absorption coefficient	0.167 mm^{-1}
F(000)	960
Crystal size	0.18 mm x 0.15 mm x 0.13 mm
Theta range for data collection	2.48 - 24.96°
Limiting indices	$-18 \leq h \leq 18, -8 \leq k \leq 1, -19 \leq l \leq 19$
Reflections collected / unique	3929 / 1680 [R(int) = 0.0204]
Completeness to theta = 24.96	99.6%
Absorption correction	ψ– scan
Refinement method	Full-matrix least-squares on F^2
Data / restraints / parameters	1680 / 0 / 194
Goodness-of-fit on F^2	1.095
Final R indices [$I>2\sigma(I)$]	R_1 = 0.0353, wR_2 = 0.0905
R indices (all data)	R_1 = 0.0495, wR_2 = 0.0976
Extinction coefficient	0.0040(7)
Largest diff. peak and hole	0.210 and -0.201 e.Å^{-3}

Table 1. Crystallographic data and structure refinement parameters for NTF

This monoclinic polymorph of NTF crystallized in I2/a space group and the asymmetric unit contains a protonated nicotinium cation and a trifluoroacetate anion. The angle between the mean plane of the pyridine ring and the mean plane of the acetate group is 48.93° where as in the triclinic form [20], it is 14° and the distance between the anion to cation is 3.134 Å which is 0.463 Å longer than the triclinic form. The structure is stabilized by N–H⋯O, O–H⋯O and C–H⋯O hydrogen bonds and hydrogen bond geometry is given in Table 2.

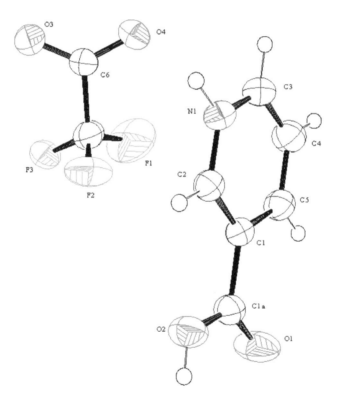

Figure 4. Structure of NTF showing 50% probability displacement ellipsoids with atom numbering scheme (for clarity, only major components of the disordered fluorine atoms are shown)

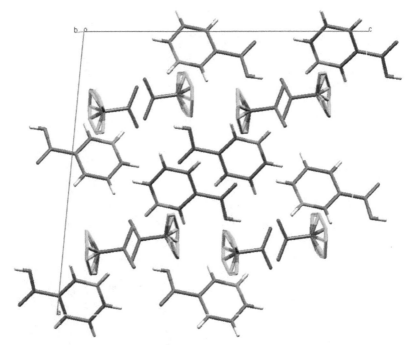

Figure 5. Molecular packing diagram of NTF viewed down the b-axis

D–H···A	d(D–H)	d(H···A)	d(D···A)	<(DHA)
N(1)–H(1N)···O(3)	0.97	1.69	2.6468	167
O(2)–H(2O)···O(4)^i	0.82	1.74	2.5460	167
C(3)–H(3)···O(1)^ii	0.93	2.44	3.3397	163
C(4)–H(4)···O(3)^iii	0.93	2.37	3.2929	174

Symmetry codes: (i) 1/2-x,y,-z, (ii) x,1/2-y,1/2+z and (iii) 1-x,-1/2+y,1/2-z.

Table 2. Hydrogen bonds geometry of NTF

CCDC No. 779179 contains the supplementary crystallographic data for this paper. These data can be obtained free of charge via www.ccdc.cam.ac.uk/data-request/cif, by e-mailing data-request@ccdc.com.ac.uk or by contacting the Cambridge crystallographic data centre, 12 Union Road, Cambridge CB21 EZ, U.K.; Fax: +44 1223 336033.

Powder X-ray diffraction study was carried out by employing SEIFERT, 2002 (DLX model) diffractometer with CuK_α (λ = 1.5405 Å) radiation using a tube voltage and current of 40 kV and 30 mA respectively. The grown crystals were finely powdered and have been subjected to powder XRD analysis. The sample was scanned over the range 10–60° at the rate of 1°/min. The indexed powder X-ray diffraction pattern of NTF is given in Figure 6. The well defined Bragg's peaks at specific 2θ angles confirmed the crystallinity of NTF.

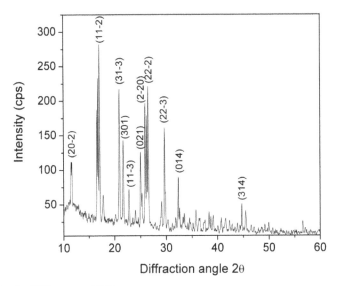

Figure 6. Powder XRD pattern of NTF

3.2. High-resolution x-ray diffraction (HRXRD) analysis

The crystalline perfection of the grown NTF single crystals was characterized by HRXRD analysis by employing a multicrystal X-ray diffractometer designed and developed at National Physical Laboratory [26]. The schematic diagram of this multicrystal X-ray diffractometer is shown in figure 7. The divergence of the X-ray beam emerging from a fine focus X-ray tube (Philips X-ray Generator; 0.4 mm × 8 mm; 2kWMo) is first reduced by a long collimator fitted with a pair of fine slit assemblies. This collimated beam is diffracted twice by two Bonse-Hart [27] type of monochromator crystals and the thus diffracted beam contains well resolved MoKα1 and MoKα2 components. The MoKα1 beam is isolated with the help of fine slit arrangement and allowed to further diffract from a third (111) Si monochromator crystal set in dispersive geometry (+, –, –). Due to dispersive configuration, though the lattice constant of the monochromator crystal and the specimen are different, the dispersion broadening in the diffraction curve of the specimen does not arise. Such an arrangement disperses the divergent part of the MoKα1 beam away from the Bragg diffraction peak and thereby gives a good collimated and monochromatic MoKα1 beam at the Bragg diffraction angle, which is used as incident or exploring beam for the specimen crystal. The dispersion phenomenon is well described by comparing the diffraction curves recorded in dispersive (+, –, –) and non-dispersive (+, –, +) configurations [28]. This arrangement improves the spectral purity ($\Delta\lambda/\lambda \ll 10^{-5}$) of the MoKα1 beam. The divergence of the exploring beam in the horizontal plane (plane of diffraction) was estimated to be ≪ 3 arc s. The specimen occupies the fourth crystal stage in symmetrical Bragg geometry for diffraction in (+, –, –, +) configuration. The specimen can be rotated about a vertical axis, which is perpendicular to the plane of diffraction, with minimum angular interval of 0.4 arc

s. The diffracted intensity is measured by using an in-house developed scintillation counter. To provide two-theta ($2\theta_B$) angular rotation to the detector (scintillation counter) corresponding to the Bragg diffraction angle (θ_B), it is coupled to the radial arm of the goniometer of the specimen stage. The rocking or diffraction curves were recorded by changing the glancing angle (angle between the incident X-ray beam and the surface of the specimen) around the Bragg diffraction peak position θ_B (taken as zero for the sake of convenience) starting from a suitable arbitrary glancing angle. The detector was kept at the same angular position $2\theta_B$ with wide opening for its slit, the so-called ω scan. This arrangement is very appropriate to record the short range order scattering caused by the defects or by the scattering from local Bragg diffractions from agglomerated point defects or due to low angle and very low angle structural grain boundaries [29].

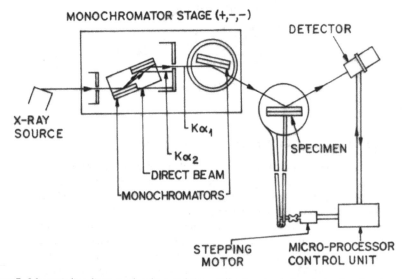

Figure 7. Schematic line diagram of multicrystal X-ray diffractometer designed, developed and fabricated at National Physical Laboratory

Before recording the diffraction curve, to remove the non-crystallized solute atoms remained on the surface of the crystal and also to ensure the surface planarity, the specimen was first lapped and chemically etched in a non preferential etchant of water and acetone mixture in 1:2 ratio.

Figure 8 shows the high-resolution X-ray diffraction curve recorded for NTF specimen crystal using $(11\bar{2})$ diffraction planes using MoKα_1 radiation. The solid line (convoluted curve) is well fitted with the experimental points represented by the filled circles. On deconvolution of the diffraction curve, it is clear that the curve contains two additional peaks, which are 18 and 28 arc s away from the main peak (at zero glancing angle). These two additional peaks correspond to two internal structural very low angle (tilt angle ≤ 1 arc min) boundaries [30] whose tilt angles (Tilt angle may be defined as the misorientation angle between the two crystalline regions on both sides of the structural grain boundary) are 18 and 28 arc s from their adjoining regions. The FWHM (full width at half maximum) of the main peak and the very low angle boundaries are respectively 22, 27 and 40 arc s. The low FWHM values and relatively low angular spread of around 200 arc s of the diffraction curve show that the crystalline perfection of the specimen is reasonably good. Thermal fluctuations or mechanical disturbances or segregation of solvent molecules during the growth process could be responsible for the observed very low angle boundaries. It may be mentioned here that such very low angle boundaries could be detected with well resolved peaks in the diffraction curve only because of the high-resolution of the multicrystal X-ray diffractometer used in the present studies.

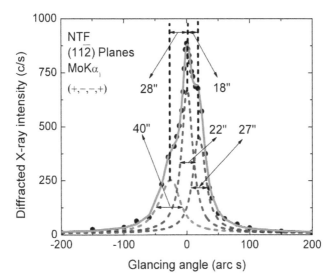

Figure 8. High-resolution X-ray diffraction curve recorded for a typical NTF single crystal specimen using $(11\bar{2})$ diffracting planes

3.3. Thermal analysis

Thermal properties of NTF were analyzed and compared with that of two nicotinium derivative crystals nicotinium oxalate (NOX) and nicotinium nitrate monohydrate (NNM). All these crystals had grown in our laboratory and belong to monoclinic system. Differential thermal analysis (DTA) and thermogravimetric analysis (TGA) of crystals were carried out simultaneously by employing TA instrument Model Q600 SDT thermal analyzer. The sample was heated at a rate of 10 °C/min in inert nitrogen atmosphere. Thermal stability of crystals was further tested using differential scanning calorimetry (DSC). DSC study was performed by using TA instrument Model Q20 in the temperature range 50–300 °C at a heating rate of 10 °C/min in inert nitrogen atmosphere and sample was placed in the Alumina crucible.

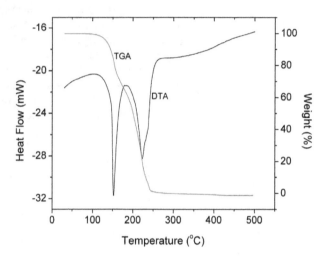

Figure 9. TGA-DTA curves of NTF

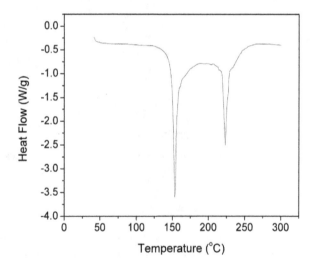

Figure 10. DSC trace of NTF

TG-DTA and DSC curves of NTF were depicted in Figures 9 and 10 respectively. In DTA curve, first endothermic peak at 152 °C is attributed to meting point of the sample, which is also evident in the DSC curve. Another important observation is that, there is no phase transition till the material melts and this enhances the temperature range for the utility of the crystal for NLO applications. The absence of water of crystallization in the molecular structure is indicated by the absence of the weight loss near 100 °C. Further there is no decomposition near the melting point [31]. This ensures the suitability of the material for possible application in lasers, where the crystals are required to withstand high temperatures. The weight loss starts around 110 °C and weight loss corresponding to decomposition of NTF was observed at 223 °C, which takes place over large temperature range (110–252 °C) where almost all the gaseous fragments like carbon dioxide and ammonia might be liberated. The TGA reveals exactly the same changes shown by DTA. The second endothermic peak in the DTA curve shows that the material is fully decomposed at 223 °C as confirmed by DSC. The small difference in the shape of the second peak may be due to the presence of impurities. The sharpness of the endothermic peaks shows good degree of crystallinity of the grown sample.

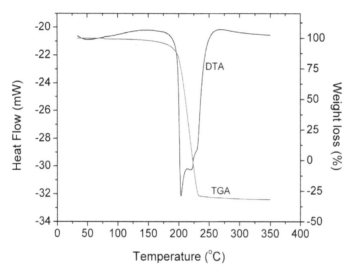

Figure 11. TGA-DTA curves of NOX

In TG-DTA trace of NOX (Figure 11), endothermic peak in DTA trace at 203 °C represents the melting point of the sample. The absence of water of crystallization in the molecular structure is indicated by the absence of the weight loss around 100 °C. TGA trace reveals that weight loss of the sample starts from this region and at 223 °C it shows complete weight loss. The shoulder peaks in DTA after the main peak corresponds to the decomposition of the material. There is no phase transition till the material melts and this enhances the temperature range for the utility of the crystal for applications. Under these conditions, phase transition means common phase transition (e.g., solid-to-liquid, liquid-to-gas etc.).

Figure 12 shows the thermogram for DTA and TGA of NNM. The compound starts to lose single molecule of water of crystallization at about 90 °C and the loss continues up to 102 °C. The weight loss in this temperature range is consistent with the weight of single molecules of water present in the crystal. The DTA curve shows a major endothermic peak, which corresponds to the melting point of NNM at 189 °C. The second weight loss take place over the temperature range 135–235 °C and almost all the compounds decomposed as its gaseous products. The second endothermic peak in the DTA curve at 230 °C attributed to the decomposition temperature of NNM. Summarized results of thermal analysis of NNM are given in Table 3.

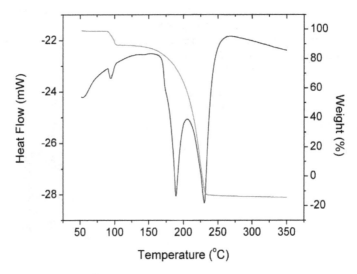

Figure 12. TGA-DTA curves of NNM

Stage	Decomposition temperature range (°C)	Decomposition steps	Weight loss (%)	
			Observed	Calculated
1	90–102	$C_6H_6NO_2.NO_3.H_2O$ $\downarrow$	8.58	8.82
2	135–235	$C_6H_6NO_2.NO_3$	91.42	91.18

Table 3. Summarized TGA and DTA results of NNM

The figure 13 shows the comparison of DSC curves of NTF with other nicotinium derivative crystals. The calculated values of enthalpy for vapourisation, melting reaction and decomposition reaction for these three materials show that enthalpy value for NTF is less than that of NNM and NOX. As the temperature increases, initially NNM loses its single molecule of water of crystallization in the range 90-102 °C. NTF melts at 152 °C, NNM at 189 °C and NOX at 203 °C respectively. Thus thermal stability of NTF is higher than that of NNM but low when compared to NOX. The low thermal stability of NNM is due to vapourisation of its water molecule.

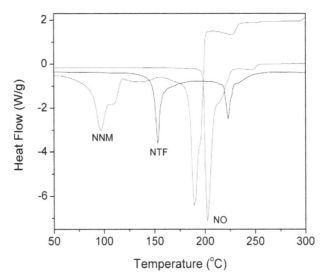

Figure 13. DSC curves of NNM, NTF and NO

3.4. FTIR studies

The FTIR spectrum was recorded using Perkin–Elmer FTIR spectrum RXI spectrometer by KBr pellet technique in the range 400–4000 cm^{-1} at room temperature (35 °C). In the FTIR spectrum of NTF (Figure 14), the strong band at 3439 cm^{-1} is attributed to stretching vibrations of O–H groups. The peak at 3086 cm^{-1} corresponds to the aromatic C–H stretching vibrations in the ring. O–H and C–C stretching vibrations are observed at 2809 cm^{-1} and 1907 cm^{-1} respectively. The existence of COO$^-$ or COOH groups in the studied crystal was deduced on the basis of vibrational spectra. It is clearly seen that the existence of COOH is illustrated by the very strong infrared band located at 1708 cm^{-1}. Asymmetric stretching vibration of COO$^-$ is observed at 1597 cm^{-1}. The peak at 1420 cm^{-1} is due to symmetric stretching vibration of COO$^-$. The C–H vibrations are occurred at 1369 cm^{-1}. C–H in plane bending vibrational modes in nicotinic acid is assigned to the frequency at 1322 cm^{-1}. It should be noted that the next band at 1194 cm^{-1} in IR spectrum is assigned to C–F stretching, which is the characteristic vibration peak of CF$_3$ group [32, 33]. The absorption at 1140 cm^{-1} is also due to the stretching type of vibrations of C–F bonds. The band at 517 cm^{-1} is assigned to C–C=O wagging. The characteristics bands, one at 836 cm^{-1} (COO$^-$ rocking), one at 746 cm^{-1} (COO$^-$ scissoring) and the third at 622 cm^{-1} (COO$^-$ wagging) are observed in the IR spectrum.

3.5. Dielectric studies

The dielectric constant is one of the basic electrical properties of solids. Dielectric properties are correlated with the electro-optic property of the crystals [34]. The capacitance (C$_{crys}$) and dielectric loss (tan δ) of NTF crystal were measured using the conventional parallel plate capacitor method with the frequency range 100 Hz to 1 MHz using the Agilent 4284A LCR meter at various temperatures ranging from 40 to 80 °C. A good quality crystal of size 5 × 5 × 2 mm³ was electroded on either side with graphite coating to make it behave like a parallel plate capacitor. The observations were made during cooling of the sample. The air capacitance (C_{air}) was also measured.

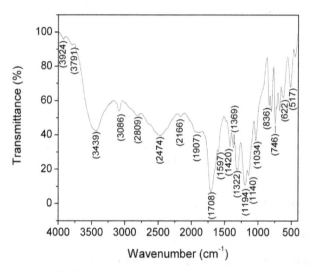

Figure 14. FTIR spectrum of NTF

The dielectric constant (ε_r) of the crystal was calculated using the following relation

$$\varepsilon_r = \frac{C_{crys}}{C_{air}} \tag{1}$$

As the crystal area was smaller than the plate area of the cell, parallel capacitance of the portion of the cell not filled with the crystal was taken into account and, consequently the above equation becomes

$$\varepsilon_r = \left(\frac{C_{crys} - C_{air}\left(1 - \dfrac{A_{crys}}{A_{air}}\right)}{C_{air}} \right) \left(\frac{A_{air}}{A_{crys}} \right) \tag{2}$$

where A_{crys} is the area of the crystal touching the electrode and A_{air} is the area of the electrode.

The variation of dielectric constant with frequency at different temperatures (Figure 15) shows that dielectric constant decreases with increasing frequency and finally it becomes almost a constant at higher frequencies for all temperatures. It is also indicates that dielectric constant increases with increase in temperature. The measurements of dielectric loss at different frequencies and temperatures show the same trend. This dielectric behavior [35] can be understood on the basis that the mechanism of polarization is similar to that of conduction process. The electronic exchange of the number of ions in the crystals gives local displacement of electrons in the direction of the applied field, which in turn gives rise to polarization namely, electronic, ionic, dipolar and space charge polarization. Space charge polarization is generally active at lower frequencies and high temperatures and indicates the perfection of the crystal. As the frequency increases, a point will be reached where the space charge cannot sustain and comply with the external field and hence the polarization decreases, giving rise to decrease in values of ε_r. At 80 °C, the dielectric constant of NTF crystal at 100 Hz is 10.851, and this value decreases to 1.955 at 1 MHz. Lowering the value of dielectric constant of the interlayer dielectric (ILD) decreases the RC delay, lowers the power consumption and reduces the crosstalk between nearby interconnects [36]. Also the materials with quite low dielectric constant lead to a small RC constant, thus permitting a higher bandwidth in the range of 10^{12} Hz for light modulation. Thus materials with low dielectric constant have considerable advantages in this context.

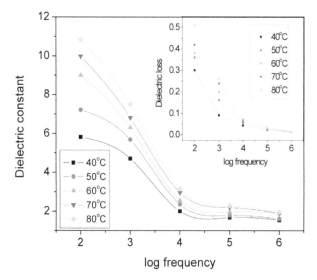

Figure 15. Plot of dielectric constant versus applied frequency. Plot of dielectric loss versus applied frequency is in inset

3.6. Mechanical hardness studies

Mechanical strength of the materials plays a key role in the device fabrication. Vickers hardness is one of the important deciding factors in selecting the processing (cutting, grinding and polishing) steps of bulk crystal in fabrication of devices based on crystals.

Microhardness measurements were done on ($\overline{1}$12) face of NTF crystal using Leitz-Wetzlar hardness tester fitted with a Vickers diamond indenter at room temperature. The Vickers microhardness number, Hv was calculated using the relation [37]:

$$H_V = 1.8544 \left(p / d^2 \right) kg / mm^2 \tag{3}$$

where p is the applied load (g) and d is the diagonal length (μm) of the indentation. The indentation time was kept at 10 s and microhardness value was taken as the average of the several impressions made. Figure 16 shows the variation of Hv as function of applied load in the range 10–100 g on ($\overline{1}$12) face of NTF crystal.

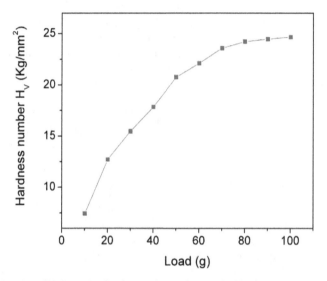

Figure 16. Variation of Vickers microhardness values versus applied load

It reveals that hardness number increases with increasing applied load. This phenomenon is known as reverse indentation size effect (RISE). When the material is deformed by the indenter, dislocations are generated near the indentation site. The major contribution to the increase in hardness is attributed to the high stress required for homogenous nucleation of dislocations in the small dislocation-free region indented [38]. The RISE can be caused by the relative predominance of nucleation and multiplication of dislocations. The other reason for RISE is that the relative predominance of the activity of either two sets of slip planes of a particular slip system or two slip systems below and above a particular load [39]. The RISE phenomenon essentially takes place in crystals which readily undergo plastic deformation [40]. The relation between load and the size of indentation can be interpreted using Meyer's law, $P = k_1 d^n$, where k_1 is a constant and n is the Meyer's number (or index). The slope of log P versus log d gives the n value and it is estimated to be 2.73. According to Onitsch [41] and

Hanneman [42], n should be between 1 and 1.6 for hard materials and above 1.6 for softer ones. Thus NTF crystal belongs to soft material category. Meyer number is a measure of the indentation size effect (ISE). For the normal ISE behaviour, the exponent n < 2. When n > 2, there is the RISE behaviour [39].

4. Conclusions

Single crystals of NTF in monoclinic system were grown by solution growth technique for the first time and its solubility and metastable zone width were determined. X-ray diffraction analysis reveals that NTF crystallizes in monoclinic system with space group $I2/a$ and unit cell parameters are a = 15.616(5) Å, b = 7.455(5) Å, c = 16.503(5) Å, and β = 95.74º. HRXRD analysis substantiates the good quality of the crystals. TG-DTA and DSC studies show that NTF melts at 152 ºC. It is observed that thermal stability of NTF is in between that of other nicotinium derivative crystals. The FTIR analysis confirms the presence of functional groups constituting NTF. Dielectric measurements indicate that NTF crystal has low values of dielectric constant and dielectric loss.

Author details

P.V. Dhanaraj
Department of Physics, Malabar Christian College, Kozhikode, India

N.P. Rajesh
Centre for Crystal Growth, SSN College of Engineering, Kalavakkam, India

5. References

[1] Zyss J, Dhenaut C, Van T C, Ledoux I (1993) Quadratic Nonlinear Susceptibility of Octupolar Chiral Ions. Chem. Phys. Lett. 206: 409-414.

[2] Russell V A, Evans C C, Li W, Ward M D (1997) Nanoporous Molecular Sandwiches: Pillared Two-Dimensional Hydrogen-bonded Networks with Adjustable Porosity. Science 276: 575-579.

[3] Matos Gomes E D, Venkataramanan V, Nogueira E, Belsley M, Proenca F, Criado A, Dianez M J, Estrada M D,Perez-Garrido S (2000) Synthesis, Crystal Growth and Characterization of a Nonlinear Optical Material–Urea L- Malic Acid. Synth. Met. 115: 225-228.

[4] Levine B F, Bethea C G (1975) Conjugated Electron Contributions to the Second Order Hyperpolarizability of Substituted Benzene Molecules. J. Chem. Phys. 63: 6-10.

[5] Bosshard Ch, Knopfle G, Pretre P, Gunter J P (1992) Second-order Polarizabilities of Nitropyridine DerivativesDetermined with Electric-Field-Induced Second-Harmonic Generation and a Solvatochromic Method: A Comparative Study. J. Appl. Phys. 71: 1594-1599.

[6] Clays K, Persoons A (1991) Hyper-Rayleigh Scattering in Solution. Phys. Rev. Lett. 66: 2980-2983.

[7] Clays K, Olbrechts G, Munters T, Persoons A, Kim O K, Choi L S (1998) Enhancement of the Molecular Hyperpolarizability by a Supramolecular Amylose–Dye Inclusion Complex, Studied by Hyper-Rayleigh Scatteringwith Fluorescence Suppression. Chem. Phys. Lett. 293: 337-342.

[8] Dhenaut C, Ledoux I, Samuel I D W, Zyss J, Bourgault M, Bozec H L (1995) Chiral Metal Complexes with Large Octupolar Optical Nonlinearities. Nature 374: 339-342.

[9] Sylla M, Giffard M, Boucher V, Illien B, Mercier N, Nguyen Phu X (1999) Nonlinear Optical Properties of Chiral Thiolates. Synth. Met. 102: 1548-1549.

[10] Ferrier J L, Gazengel J, Nguyen Phu X, Rivoire G (1984) Backscattering in the Picosecond Range: an Optical Triggered Switching Effect. Opt. Commun. 51 (4): 285-288.

[11] Somac M, Somac A, Davies B L, Humphery M G, Wong M S (2002) Third-Order Optical Nonlinearities of Oligomers, Dendrimers and Polymers Derived from Solution Z-Scan Studies. Opt. Mater. 21: 485–488.

[12] Natarajan L V, Sutherland R L, Tondiaglia V P, Bunning T J, Adams W W (1996) Electro-Optical Switching Characteristics of Volume Holograms in Polymer Dispersed Liquid Crystals. J. Nonlinear Opt. Phys. Mater. 5: 89- 98.

[13] Pecaut J, Bagieu-Beucher M (1993) 2–Amino–5–nitropyridiniummonohydrogenphos phite. Acta Cryst. C49: 834-837.

[14] Ravindra H J, John Kiran A, Dharmaprakash S M, Satheesh Rai N, Chandrasekharan K, Kalluraya B, Rotermund F (2008) Growth and characterization of an efficient nonlinear optical D–π–A–π–D type chalcone single crystal. J. Cryst. Growth 310: 4169-4176.

[15] Gielen M, Kholufi A E, Biesemans M, Willem R (1992) (2-Methylthio-3-Pyridinecarboxylato)-diethyltin and -di- n-butyltin Compounds: Synthesis, Spectroscopic Characterization and in vitro Antitumour Activity. X-ray Crystal Structure of bis[diethyl(2-methylthio-3-Pyridinecarboxylato)tin] oxide and of diethyltin bis(2-methylthio-3- pyridinecarboxylate), Polyhedron. 11: 1861-1868.

[16] Athimoolam S, Anitha K, Rajaram R K (2005) Nicotinium dihydrogenphosphate. Acta Cryst.E61: o2553–o2555.

[17] Athimoolam S, Rajaram R K (2005) Bis(nicotinic acid) hydrogen perchlorate. Acta Cryst. E61:o2674–o2676.

[18] Athimoolam S, Rajaram R K (2005) Dinicotinium sulfate. Acta Cryst. E61: o2764–o2767.

[19] Gao S, Liu J W, Huo L H, Sun Z Z, Gao J S, Ng S W (2004) Catena-Poly[[diaquabis(2-chloronicotinato-κ ^{2}O,O')cadmium(II)]-μ-2-chloronicotinato-κ^{3}O,O':N]. Acta Cryst. C 60: m363-m365.

[20] Athimoolam S, Natarajan, S (2007) Nicotinium trifluoroacetate. Acta Cryst. E 63: 2656-2657.

[21] Nyvlt J, Rychly R, Gottfried J, Wurzelova J (1970) Metastable Zone-Width of Some Aqueous Solutions. J. Cryst. Growth 6: 151-162.

[22] Sangwal K (1989) On The Estimation of Surface Entropy Factor, Interfacial Tension, Dissolution Enthalpy and Metastable Zone-Width for Substances Crystallizing from Solution. J. Cryst. Growth 97: 393-405.

[23] Harms K, Wocadlo S, XCAD4, University of Marburg, Germany, 1995.

[24] North A C T, Phillips D C, Mathews F S (1968) A semi-empirical method of absorption correction. Acta Cryst. A24: 351-359.

[25] SHELXTL/PC Version 6.10 Madison, WI: Bruker AXS Inc., 2000.

[26] Lal K, Bhagavannarayana G (1989) A High-Resolution Diffuse X-Ray Scattering Study of Defects in Dislocation-Free Silicon Crystals Grown by the Float-Zone Method and Comparison with Czochralski-Grown Crystals. J. Appl. Cryst. 22: 209-215.

[27] Bonse U, Hart M (1965) Tailless X–ray Single Crystal Reflection Curves Obtained by Multiple Reflection. Appl. Phys. Lett. 7: 238-240.

[28] Bhagavannarayana G (1994) High Resolution X-Ray Diffraction Study of As-Grown and $BF_2{}^+$ Implanted Silicon Single Crystals, Ph. D. Thesis, University of Delhi, Delhi, India.

[29] Bhagavannarayana G, Kushwaha S K (2010) Enhancement of SHG Efficiency by Urea Doping in ZTS Single Crystals and its Correlation with Crystalline Perfection as Revealed by Kurtz Powder and High-Resolution X-Ray Diffraction Methods. J. Appl. Cryst. 43: 154-162.

[30] Bhagavannarayana G, Ananthamurthy R V, Budakoti G C, Kumar B, Bartwal K S (2005) A Study of the Effect of Annealing on Fe-Doped $LiNbO_3$ by HRXRD, XRT and FT-IR. J. Appl. Cryst. 38: 768-771.

[31] Willard, Merritt, Dean, Settle (1986) Instrumental Methods of Analyses, First Indian Edition: CBS, Delhi.

[32] Fuson N, Josien M L, Jones E A, Lawson J R (1952) Infrared and Raman Spectroscopy Studies of Light and Heavy Trifluoroacetic Acids. J. Chem. Phys. 20: 1627-1635.

[33] Takeda Y, Suzuki H, Notsu K, Sugimoto W, Sugahara Y (2006) Preparation of a Novel Organic Derivative of the Layered Perovskite Bearing $HLaNb_2O_7 \cdot nH_2O$ Interlayer Surface Trifluoroacetate Groups. Mat. Res. Bull. 41: 834-841.

[34] Aithal P S, Nagaraja H S, Mohan Rao P, Avasti D K, Sarma A (1997) Effect of high energy ion irradiation on electrical and optical properties of organic nonlinear optical crystals. Vacuum 48: 991-994.

[35] Anderson J C (1964) Dielectrics, Chapman and Hall.

[36] Hatton B D, Landskron K, Hunks W J, Bennett M R, Shukaris D, Pervoic D D, Ozin G A (2006) Materials Chemistry for Low k-Materials. Mater. Today. 9: 22-31.

[37] Mott B W (1956) Microindentation Hardness Testing, Butterworths, London.

[38] Kunjomana A G, Chandrasekharan K A (2005) Microhardness Studies of GaTe Whiskers. Cryst. Res. Technol. 40: 782-785.

[39] Sangwal K (2000) On the Reverse Indentation Size Effect and Microhardness Measurement of Solids. Mater. Chem. Phys. 63: 145-152.

[40] Li H, Han Y H, Bradt R C (1994) Knoop Microhardness of Single Crystal Sulphur. J. Mater. Sci. 29: 5641-5645.

[41] Onitsch E M (1947) Mikroscopia 2: 131-134.
[42] Hanneman M (1941) Metall. Manch. 23: 135-139.

Application of Isothermal Titration Calorimetry for Analysis of Proteins and DNA

Isothermal Titration Calorimetry: Thermodynamic Analysis of the Binding Thermograms of Molecular Recognition Events by Using Equilibrium Models

Jose C. Martinez, Javier Murciano-Calles, Eva S. Cobos, Manuel Iglesias-Bexiga, Irene Luque and Javier Ruiz-Sanz

Additional information is available at the end of the chapter

1. Introduction

The revolution achieved during the last decade in the fields of genomics and proteomics has shown the need of going in-depth into the structural, dynamic, energetic and functional knowledge of biological macromolecules, mainly proteins and nucleic acids. Of special interest is the study of the molecular recognition between such kind of molecules or between them and other biological molecules, for example, natural metabolites or designed drugs to alter their functionalities.

Isothermal titration calorimetry (ITC) is a technique that directly measures the heat exchange accompanying a chemical or biochemical reaction. It is the ideal technique for the investigation of the energetics of ligand binding to biological macromolecules because it provides a complete thermodynamic characterization of the macromolecule-ligand interactions, allowing for the measurement of the binding affinity as well as of the changes in enthalpy and entropy of the process. The nature (enthalpic or entropic) and magnitude of the forces directing the interaction are very important factors to be considered in the design of ligands with specific characteristics. Additionally, the heat capacity of binding can be determined by carrying out titration experiments at different temperatures.

From the diverse methodologies that can be applied in the research of binding processes, ITC presents a series of advantages and possibilities, and as such it is considered a very powerful tool. Precluding the structural interpretation, the direct determination of binding thermodynamic parameters becomes necessary to describe the energetic aspects of the binding and, thus, to define and to rationalize macromolecular recognition. Nevertheless, although calorimetry has been widely used as an experimental resource, it has not always been interpreted correctly, mainly due to the difficulty found in extracting thermodynamic information from experimental data. Thus, the rigorous analysis of ITC thermograms should be done under the assumption of theoretical models, able to describe the most significant stages present during the binding process and which application would give rise to valuable thermodynamic information for each of such stages.

2. General aspects of binding equilibrium

Through this Chapter we are going to scrutinise the use of ITC in the study of binding equilibrium processes, as well as how to design and perform the experiments and the correct way to handle the data and achieve the corresponding fit to the proper equilibrium models. Nevertheless, prior to focusing on the different ITC aspects, we will describe briefly some basic features of binding equilibrium, for which it is crucial to introduce some basic concepts and equilibrium formulas.

2.1. Basic concepts

Apart from the capacity of self-copying, biomolecules are characterized by their ability to specifically interact with other molecules within the cell, which defines their biological functionality. Many of the biochemical processes occurring in living systems are based on, or regulated by, binding interactions between biological macromolecules or with other small molecules. Examples of interactions between macromolecules can be found out in interactions between polypeptide chains to form the quaternary structure of multi-subunit proteins, in the close association of protein and RNA molecules in the ribosome, in the binding of transcription regulators to DNA, protein-protein interactions in many signalling cascades, etc. Besides, many biological macromolecules bind small molecules, for example, enzymes that bind substrates and effector molecules, or proteins that bind metabolites in order to transport or store them. Signalling transmission is also based on interactions, as those of hormones with membrane receptors. Additionally, some of the regulation pathways of the transcription and replication of nucleic acids involve the change of their conformations induced by binding of metallic ions.

The interactions that can take place under different backgrounds and contexts from a physico-chemical point of view, can be summarized into three different types: i) at *equilibrium*, ii) at *steady-state conditions*, and iii) at the *transition between different steady-state conditions*. In this Chapter, we will direct attention to the first case, the binding equilibrium process between a biological macromolecule (such as a protein or a nucleic acid) and a small molecule, called a *ligand*, occurring by *specific* interactions, that is, the ligand (L) binds at specific sites of the macromolecule (M). The establishment of such specific interactions is crucial for the correct functioning of the cell, as happens in the most of biological processes, where one or more macromolecule-ligand (ML) interactions are involved, determining and regulating the biological function.

All these ML interactions present some common features:

- Binding of the ligand involves a non-covalent *reversible* interaction to a specific region of the macromolecule, called the *binding site*, usually situated at its surface or close to it.
- The ligand binding process may induce conformational changes that modify the activity of the macromolecule; this phenomenon is known as *alosterism*.
- When the macromolecule has more than one binding site, the binding of one ligand to one of the sites may change the affinity of the ligand for the rest of binding sites; this feature is known as *co-operativity* and is closely related to the alosterism phenomenon.
- In some cases, the binding process can result in a change in the aggregation state of the molecules (*polisterism*) or, even, give rise to a new phase in the system (*poliphasic* processes). These two aspects are not within the scope of this Chapter.

The correct characterization of the binding process requires some experimental work in order to determine a variety of parameters such as:

- **Number of binding sites** per macromolecule for a defined ligand, n. The numeric value can be one or higher, sites can be identical or different in terms of affinity into the same macromolecule.
- **Binding parameter,** $\bar{v}$. Represents the moles of bound ligand by each mole of macromolecule. It ranges between zero and the number of binding sites, n.
- **Saturation fraction, θ.** The fraction of the total number of sites of the macromolecule occupied by ligand molecules, which ranges from zero (no occupancy) to one (all sites occupied). It can be easily deduced that $\bar{v} = n \cdot \theta$.
- **Binding affinity** of the ligand to the macromolecule, expressed by means of the equilibrium binding constant, K_b, or the corresponding Gibbs energy change, $\Delta G_b = -RT \cdot lnK_b$. As mentioned above, ITC experiments provide a complete thermodynamic characterization of the macromolecule-ligand interactions, allowing the determination of the binding affinity as well as the changes in enthalpy, ΔH_b, and entropy, ΔS_b, of the binding process, where $\Delta G_b = \Delta H_b - T \cdot \Delta S_b$.
- $\bar{v}/(n-\bar{v})$ is the **relationship between occupied ($\bar{v}$) and empty (n-$\bar{v}$) sites** in the macromolecule.

Thus, binding studies can provide the answer to some fundamental questions related to the functional aspects of biological macromolecules, such as, for example: How many binding sites in the macromolecule for a defined ligand exist? What is the affinity of the ligand for each binding site? Is there any dependency or inter-connection among the sites? Can affinity be modulated by the proper ligand molecule or by any other metabolites?

The experimental data is ideally expressed in terms of changes in the binding parameter, $\bar{v}$, as a function of the free ligand concentration in solution, $[L]$. In practice, it is necessary to move along the whole equilibrium process, starting usually from a solution containing the free macromolecule where the ligand solution is added progressively until the saturation of all sites is achieved. During this titration process, we should measure the binding parameter, generally by using spectroscopic or calorimetric probes. This kind of approach allows us to know the total concentration of both macromolecule, $[M]_T$, and ligand, $[L]_T$, in the solution.

2.2. The Adair's equation

This equation defines the type of equilibrium that can be established between a macromolecule and its ligand upon binding.

2.2.1. Binding to one site

In order to establish how the equilibrium constants can be determined from experimental data we are going to develop the simplest binding process, described by the binding of a ligand to a macromolecule which has only one binding site. It can be expressed as following:

$$M + L \xleftrightarrow{K_b} ML \qquad\qquad K_b = \frac{[ML]}{[M][L]} \qquad\qquad (1)$$

Although the thermodynamic equilibrium constant that characterizes the binding process, K_b, must be expressed as a function of the activities of the different species present at equilibrium, it is usual to use concentrations instead of this, as experimental data contains larger errors than the ones derived from this approximation.

As we have stated previously, a very useful parameter obtained by experimental data is the *binding parameter*, $\bar{v}$, defined as the average of ligand molecules that are bound per macromolecule, its range from 0 to n (number of binding sites per macromolecule). Mathematically it can be defined as:

$$\bar{v} = \frac{[L]_b}{[M]_T} = \frac{[ML]}{[M]+[ML]} = \frac{K_b[L]}{1+K_b[L]} \qquad\qquad (2)$$

The representation of $\bar{v}$ *versus [L]* (free ligand concentration) gives the so called *binding curve*. As shown in Figure 1, our simple example corresponds to a hyperbolic curve trending asymptotically to the number of sites n ($n=1$ in this case), because saturation conditions are reached as the free ligand concentration increases.

The value of the equilibrium constant, K_b, can be determined from the non-linear fitting of the experimental data, represented in the binding curve, to equation 2. In the case of a single binding site, it is also possible to convert equation 2 into a variety of linear equations to obtain the K_b value from the corresponding linear regression.

Such linear representations (Figure 1) can be easily deduced from the previous equations and are named as follows:

Double reciprocal representation:

$$\frac{1}{\bar{v}} = 1 + \frac{1}{K_b[L]} \qquad\qquad (3)$$

Hill representation:

$$\log\left(\frac{\bar{v}}{1-\bar{v}}\right) = \log\left(K_b[L]\right) \tag{4}$$

• Scatchard representation:

$$\frac{\bar{v}}{[L]} = K_b - K_b\bar{v} \tag{5}$$

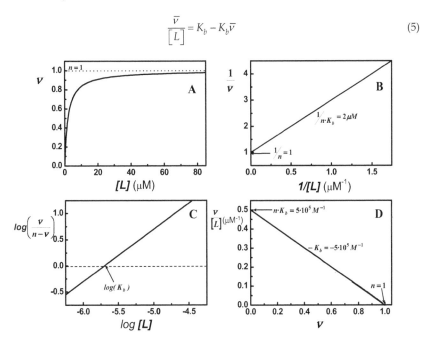

Figure 1. Binding to one site
Different representations of simulated data for the binding equilibrium of a ligand to a macromolecule with a single binding site with $K_b = 5\cdot10^5$ M^{-1}. (A) Binding curve. (B) Double reciprocal representation. (C) Hill representation. (D) Scatchard representation.

Another accessible parameter from binding experiments is the *saturation fraction*, θ, defined as the fraction of binding sites occupied by ligand. It is related to the binding parameter by the expression: $\theta = \bar{v}/n$

Depending on the techniques used to obtain the experimental data of the binding equilibrium processes, it is possible to determine either the binding parameter, $\bar{v}$, or the saturation fraction, θ, and then, to develop the subsequent analysis to determine the rest of parameters of interest.

The most appropriate technique to determine $\bar{v}$ is ITC [1, 2]. It can also be achieved by equilibrium dialysis, which allows the calculation of the concentrations of the free ligand in equilibrium with the different species of the macromolecule (free and bound), in order to construct directly the binding curve. Although this technique provides the complete set of experimental data required, it requires extensive work and also needs large amounts of sample.

The techniques that allow the determination of θ values are based on detecting physical change occurring in the macromolecule or in the ligand during the binding process. Such physical change has to be a linear function, as the ligand is bound to the macromolecule. Furthermore, if the macromolecule presents several binding sites, the change must be the same for all of them or, at least, the change relationship between the binding sites must be known. Depending on the nature of the physical change, different techniques may be used: UV-visible spectroscopy, fluorescence, circular dicroism, nuclear magnetic resonance spectroscopy, etc [1, 2].

2.2.2. Binding to two equivalent and independent sites

2.2.2.1. Macroscopic formulae

Here, we are going to describe the formulae of the equilibrium processes corresponding to the binding of a ligand to a macromolecule with two binding sites. In this stage, we will focus on the simplest situation, where both sites are equal in affinity and independent, *i.e.*, not influencing each other upon ligand binding. Binding schemes where these basic assumptions do not occur can be useful to describe cooperative interactions and will be described later on in this Chapter. To obtain the binding parameters we can use elementary thermodynamics for the simplest non-cooperative cases, but as the cases turn more complex, this formulae becomes very laborious and a more general formulae is needed.

In order to strengthen the binding concepts, we will start by applying the classical formulae to this simple case, before the description of the general formulae introduced in the biochemical field by Wyman [3], which is useful for the formulae of more complicated binding schemes. Thus, in the case of a macromolecule with two equivalent and independent binding sites, the description of the formulae from a **macroscopic** point of view can be developed in two different but equivalent ways:

- Stage formulae: A first equilibrium stage is considered, where M binds to one ligand molecule, L, followed by a second stage where a second L molecule binds to M, achieving saturation. These two stages are characterized by their corresponding equilibrium constants as follows:

$$M + L \xleftrightarrow{\;K_{b1}\;} ML \qquad\qquad K_{b1} = \frac{[ML]}{[M][L]} \qquad K_{b2} = \frac{[ML_2]}{[ML][L]} \qquad (6)$$

$$ML + L \xleftrightarrow{\;K_{b2}\;} ML_2$$

- Global formulae: The equilibriums take place between both the free and the bound M species to either one or two ligand molecules. The equilibrium constants for this formulae are related with the ones of the previous formulae as follows:

$$M + L \xleftrightarrow{\;\beta_1\;} ML \qquad\qquad \beta_1 = \frac{[ML]}{[M][L]} = K_{b1} \qquad \beta_2 = \frac{[ML_2]}{[M][L]^2} = K_{b1} \cdot K_{b2} \qquad (7)$$

$$M + 2L \xleftrightarrow{\;\beta_2\;} ML_2$$

The binding equilibrium constants K_b and β are named as *macroscopic constants*.

In this case, where the macromolecule has two equivalent and independent binding sites, the binding parameter, $\bar{v}$, can be calculated as:

$$\bar{v} = \frac{[ML] + 2[ML_2]}{[M] + [ML] + [ML_2]} = \frac{K_{b1}[L] + 2K_{b1}K_{b2}[L]^2}{1 + K_{b1}[L] + K_{b1}K_{b2}[L]^2} = \frac{\beta_1[L] + 2\beta_2[L]^2}{1 + \beta_1[L] + \beta_2[L]^2} \qquad (8)$$

2.2.2.2. Microscopic formulae

When more than one binding site exists , the binding process can be described using a **microscopic formulae**, that is, distinguishing each binding site. Thus, as it is shown in Figure 2, the first ligand molecule can bind to binding site 1 of M, being the equilibrium process characterized by the microscopic binding constant k_1; or to the binding site 2 and then, characterized by k_2. The second ligand molecule will bind to the free binding site, and the equilibrium will be characterized by k_3 if the free site is the number 2 or k_4 if it is the number 1.

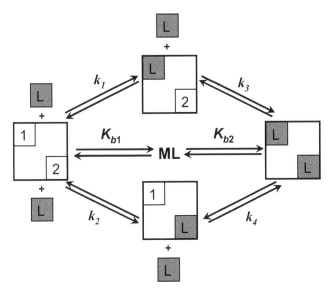

Schematic representation to distinguish between the macroscopic and microscopic formulae for the binding equilibriums of a general ligand L to a macromolecule with two binding sites.

Figure 2. Binding to two sites

For the case we are explaining, the meaning of having equivalent binding sites is that $k_1 = k_2$ and $k_3 = k_4$. Also, independent binding sites imply that $k_1 = k_4$ and $k_2 = k_3$. So, in binding processes where all binding sites are equivalent and independent, all microscopic equilibrium constants will be identical. The relations between macroscopic, K_b, and microscopic, k, equilibrium constants are: $K_{b1} = 2k$ and $K_{b2} = \dfrac{k}{2}$. Thus, although microscopic constants are identical, the macroscopic ones are different due to statistical factors. By using the microscopic constants instead of the macroscopic ones, a simpler expression for the binding parameter than that given in equation 8 is obtained:

$$\bar{v} = \frac{2k[L]}{1 + k[L]} \qquad (9)$$

Therefore, it is interesting to know the relationships between the equilibrium constants obtained using the different formulae; such relationships between microscopic and macroscopic constants may allow it to be deduced whether the binding sites are independent or not. Meanwhile, the use of microscopic constants will simplify the equation of the binding parameter which, as mentioned above, is the experimentally accessible parameter, besides the fraction saturation.

2.2.3. Binding to n equivalent and independent sites

We can obtain the relationship between the different types of binding constants for the general case of a macromolecule having n equivalent and independent binding sites. Firstly, we proceed to apply the macroscopic formulae in its two ways, which are summarized in the next scheme:

$$
\begin{array}{cc}
\textit{Stage formulation} & \textit{Global formulation} \\[6pt]
\text{M+L} \;\xrightleftharpoons{K_{b1}}\; \text{ML} & \text{M+L} \;\xrightleftharpoons{\beta_1}\; \text{ML} \\[6pt]
\text{ML+L} \;\xrightleftharpoons{K_{b2}}\; \text{ML} & \text{M+2L} \;\xrightleftharpoons{\beta_2}\; \text{ML} \\
\vdots & \vdots \\
\text{ML}_{i-1} \;\xrightleftharpoons{K_{bi}}\; \text{ML}_i & \text{M+iL} \;\xrightleftharpoons{\beta_i}\; \text{ML}_i \\
\vdots & \vdots \\
\text{ML}_{n-1} \;\xrightleftharpoons{K_{bn}}\; \text{ML}_n & \text{M+nL} \;\xrightleftharpoons{\beta_n}\; \text{ML}_n
\end{array} \qquad (10)
$$

Looking at the equations defining the macroscopic binding parameter (equations 2 and 8) we can easily deduce that, in the case of having n equivalent and independent binding sites, this variable can be written in general as

$$\bar{v} = \frac{[L]_b}{[M]_T} = \frac{\displaystyle\sum_{i=1}^{n} i[ML_i]}{\displaystyle\sum_{i=0}^{n} [ML_i]} \qquad (11)$$

where ML_i refers to the macromolecule with i bound ligand molecules. For a general stage i we can obtain the relation between both macroscopic constants, considering that the concentration of ML_i may be expressed as:

$$\left[ML_i \right] = K_i \left[ML_{i-1} \right] \cdot \left[L \right] = K_i \cdot K_{i-1} \cdot \left[ML_{i-2} \right] \cdot \left[L \right]^2 = K_i \cdot K_{i-1} \cdot \ldots \cdot K_1 \cdot \left[M \right] \cdot \left[L \right]^i \tag{12}$$

So:

$$\beta_i = \frac{\left[ML_i \right]}{\left[M \right]\left[L \right]^i} = K_i K_{i-1} \cdots K_1 = \prod_{j=1}^{i} K_j \tag{13}$$

Taking into account the equation 11, the binding parameter, expressed in terms of the macroscopic constants, is:

$$\bar{v} = \frac{\displaystyle\sum_{i=1}^{n} i \cdot \beta_i \left[L \right]^i}{\displaystyle\sum_{i=0}^{n} \beta_i \left[L \right]^i} \tag{14}$$

This equation is known as *Adair's general equation*, being the denominator the so called *binding polynomial*.

As equation 14 presents a very high number of macroscopic constants, it is interesting to deduce the relation between microscopic and macroscopic equilibrium constants, to obtain a more simple expression for the binding parameter. Firstly, it is necessary to know the number of possible microscopic states of each macroscopic species. Thus, for ML_i species it will be the number of different ways to arrange i ligands into n binding sites, which corresponds to the combinatorial of n elements taken in groups of i. Since all binding sites are equivalent and independent, all the possible microscopic forms for any macroscopic species are equally probable and, therefore, they will be at the same concentration; then, the concentration of the macroscopic ML_i species expressed as the concentration of its microscopic forms is:

$$\left[ML_i \right] = \frac{n!}{(n-i)!\,i!}\left[ML_i \right]_{micro} \tag{15}$$

So the microscopic equilibrium constant, k, will be:

$$k = \frac{\left[ML_i \right]_{micro}}{\left[ML_{i-1} \right]_{micro}\left[L \right]} = \frac{(n-i)!\,i!\left[ML_i \right]}{(n-i+1)!(i-1)!\left[ML_{i-1} \right]\left[L \right]} = \frac{i}{n-i+1} K_{bi} \tag{16}$$

The macroscopic constant β_i is obtained from equations 13 and 16 as

$$\beta_i = \frac{n!}{(n-i)!\,i!} k^i \tag{17}$$

The binding parameter expressed in terms of microscopic constants can be obtained from equations 14 and 17 as

$$\bar{v} = \frac{\sum\limits_{i=1}^{n} i \frac{n!}{(n-i)!i!} k^i [L]^i}{\sum\limits_{i=0}^{n} \frac{n!}{(n-i)!i!} k^i [L]^i} = \frac{nk[L](1+k[L])^{n-1}}{(1+k[L])^n} = \frac{nk[L]}{1+k[L]} \tag{18}$$

where the resulting expression has been obtained by taking into account the binomial theorem. Once again, the binding parameter, expressed in terms of microscopic constants, gives a quite simple expression with a reduced number of fitting parameters, since a single k value is always expected for all microscopic binding constants in the case of a binding process where binding sites are equivalent and independent.

2.3. A general formulae for non-cooperative binding. The binding polynomial

At this point, we have explained the simplest cases of binding equilibrium, where binding sites are equivalent and independent. When more complex schemes are considered, the use of classical thermodynamic formulae to obtain the binding parameter turns complicated and laborious, as was mentioned previously. Thus, it is more convenient to use a general formulae which will allow obtaining $\bar{v}$ ($[L]$) expressions, systematically and independently of the complexity of the case in study. This general formulae is based on the construction of a function, which may be described as the macroscopic analogue of the grand canonical partition function from statistical thermodynamics. This function was introduced in the biochemical field by Wyman [3] and then, applied to ligand binding [4-7].

In order to apply this general formulae to ligand binding systems, firstly, it is necessary to construct the *partition function*, and then, apply it to the system under study. Let us explain briefly the steps to obtain the partition function:

- Identify the different energetic accessible states of the system: in a binding process it would be the different species of the macromolecule (free and bound) in equilibrium with the ligand.
- Determine the energy of each accessible state: it would be equivalent to specify the equilibrium constants given by the mass action law.
- Choose a reference state, that is, a reference specie: it can be chosen any of the identified ones, though is preferable to choose the state (species) with the lowest energy; for a binding process is the free macromolecule (M).
- Calculate the statistical weight of each state (species) with respect to the reference one. The statistical weight for a state i, W_i, is

$$W_i = D * \exp\left(-\Delta E_i \Big/ RT\right) \tag{19}$$

where D is the degeneration of each state, and $\Delta E_i = E_i - E_{ref}$

For a binding process the statistical weight of the specie ML_i is

$$W_i = [ML_i]/[M] = \beta_i [L]^i \tag{20}$$

Construct the partition function, Z, as the sum of the statistical weights of all accessible states (or species): $Z = \sum_i W_i$. For binding processes the partition function is expressed as

$$Z = \sum_{i=0}^{n} \frac{[ML_i]}{[M]} = \sum_{i=0}^{n} \beta_i [L]^i \tag{21}$$

Based on this formalism it is possible to easily obtain interesting expressions such as

Probability of each accessible state, P_i, which is the fraction of each species at equilibrium in the binding process

$$P_i = \frac{W_i}{Z} = \frac{\beta_i [L]^i}{\sum_{i=0}^{n} \beta_i [L]^i} \tag{22}$$

- Average quantities of interest for the system, that is, measured values of any magnitude. For a magnitude a

$$\langle a \rangle = \sum_{i=0}^{n} a_i \cdot P_i \tag{23}$$

where a_i corresponds to the value of the magnitude a for the state i (specie i).

Thus, the binding parameter, $\bar{v}$, corresponds to the average of ligand bound to the macromolecule and can be calculated as follows

$$\bar{v} = \langle i \rangle = \sum_{i=0}^{n} i \cdot \frac{W_i}{Z} = \sum_{i=0}^{n} i \cdot \frac{\beta_i [L]^i}{\sum_{i=0}^{n} \beta_i [L]^i} = \frac{\sum_{i=0}^{n} i \cdot \beta_i [L]^i}{\sum_{i=0}^{n} \beta_i [L]^i} \tag{24}$$

Of course, the expression obtained is the Adair's equation (equation 14) and the denominator, which corresponds to the binding polynomial, can be identified with the partition function.

There is also a direct way to calculate the binding parameter, based on the calculation of the partial derivative of $\ln Z$ in respect to $\ln[L]$ at constant P and T, as is shown below:

$$\bar{v} = \left(\frac{\partial(\ln Z)}{\partial(\ln[L])} \right)_{P,T} = \frac{[L]}{Z} \left(\frac{\partial Z}{\partial[L]} \right)_{P,T} = \frac{[L]}{Z} \left(\frac{\partial \left(\sum_{i=0}^{n} \beta_i [L]^i \right)}{\partial[L]} \right)_{P,T} = \frac{\sum_{i=0}^{n} i \cdot \beta_i [L]^i}{\sum_{i=0}^{n} \beta_i [L]^i} \tag{25}$$

If we apply this general formulae to the case of the binding of a ligand to a macromolecule with n equivalent and independent binding sites, we can obtain the same expression than that given in equation 18. It is interesting to note that, since the binding sites are independent, the partition function corresponds to the product of the partition sub-function for each binding site. Therefore, the binding parameter can also be expressed as the sum of the binding parameter for each binding site. Additionally, if the sites are equivalent, such parameters will be equal to n times the value of the binding parameter obtained for one of the sites.

In the case of a macromolecule with n different and independent binding sites, the partition function can also be expressed as the product of the partition sub-functions for each binding site, though as the sites are different these partition sub-functions will not be equivalent. Consequently, the binding parameter will be the sum of the binding parameter corresponding to each site.

2.4. Experimental analysis of binding equilibriums to independent sites

Prior to describing cooperative phenomena, in this Section we will describe how the analysis of the different graphical representations mentioned in Section 2.2.1 may help to rationalize the experimental data to get information about, for example, the existence of different kinds of binding sites for the ligand and how the equilibrium constants describing ligand binding can be estimated.

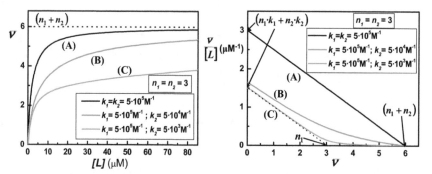

Simulation examples for the binding equilibrium of a ligand to a macromolecule with one or two different classes of sites, to show the differences in the binding curves (left panel) and in the Scatchard representation (right panel). Curve A corresponds to the ligand binding to six equivalent and independent sites, with a microscopic constant of $5\cdot10^5$ M^{-1}. Curves B and C correspond to the binding to two different kinds of sites, each class with 3 sites: for curve B the ratio between microscopic constants of the two binding site classes is $k_1=10k_2$; for curve C the ratio is $k_1=100k_2$.

Figure 3. Binding to independent sites

The easiest representation of experimental data is the binding curve (Figures 1 and 3). Simulations carried out with the equations described above, by using the different $\bar{v}$ ([L]) expressions obtained in the earlier sections indicate that a hyperbolic shape of this curve will represent a binding process corresponding to a macromolecule with equivalent and independent sites, being n estimated from the asymptotic value of the graph (curve A). When the macromolecule displays different kinds of sites, characterized by different values of microscopic constants, the shape of the curve changes, becoming more difficult to distinguish from the former simpler case when such values become similar (curves B and C). The shape will be the equivalent to the sum of two or more hyperbolic binding curves with two or more different k values.

The Scatchard representation would be more helpful to distinguish between equal or different kinds of independent sites (Figure 3). From the intersection with the X-axis the number of binding sites can be easily determined when sites are equal (curve A). In the case of different kind of sites n estimation is difficult, although we can obtain the number of sites of the highest affinity and the respective k value from the extrapolation of the initial linear tendency (see figure).

The Hill representation is fundamentally informative to distinguish between independent (non-cooperative) and dependent (cooperative) sites in the macromolecule, and will be analyzed in detail into the next Section.

2.5. Experimental analysis of binding cooperativity (non-independent sites)

Up to now we have referred only to binding processes where all binding sites are independent. It is time to introduce the effect of binding cooperativity, which means that the interaction of the ligand with one of the sites of the macromolecule produces an alteration of the affinity that the other sites have for such ligand. We can distinguish between *positive cooperativity*, when the binding of a ligand increases the affinity of the rest of binding sites, and *negative cooperativity*, when such affinity is decreased. These changes in affinity are usually related to conformational changes in the macromolecule, that is, what was referred to at the beginning of the chapter as *alosterism*. From a practical point of view, cooperativity can be viewed as a way to regulate the biological activity as a function of ligand concentration.

Although the Adair equation is still a valid approach, as we described in Section 2.4, it could be difficult to distinguish among the different equilibriums and, in addition, it may contain an excessive number of equilibrium constants to be estimated from fitting. Thus, different strategies have been described to analyze cooperativity, as will be explained into the next Section. Prior to this description, let us first describe how to determine cooperativity from analysis of experimental data.

The binding curve reveals an S-shape when cooperativity is positive. From the Scatchard representation positive cooperativity can also be easily discernible from any scheme of independent binding sites, since a concave shape of the experimental data is revealed (Figure 4). However, negative cooperativity can be confused with the scheme of different and independent sites (compared to Figures 3 and 4). The Hill representation is the most useful to distinguish between dependent and independent types of binding sites. In Figure 4 both situations have been simulated. When a non-cooperative behaviour is revealed, a single straight line with a slope equal to one is obtained, while when a positive cooperative behaviour occurs, the Hill analysis shows an increase of the slope at the central region (S-shape). The decrease of this region will indicate negative cooperativity. The slope of this region is known as the *Hill coefficient, n_H*. The explanation is as follows:

At very low ligand concentrations, the binding occurs statistically at different sites allocated in different macromolecule units, all of them free of ligand. Thus, at this stage cooperativity phenomena are not revealed experimentally. This "non-cooperative" initiation of the binding process results in a straight line with slope equal to one. Thus, equation 4 converts into

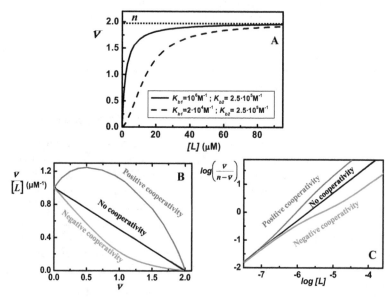

Graph A shows some simulations of the binding curve for the ligand binding to a macromolecule with two binding sites, with a microscopic constant of $k=5\cdot10^5$ M^{-1}: solid line curve corresponds to equal and independent binding sites, and dashed line curve to equal sites showing positive cooperativity (the binding of a ligand increases fifty times the affinity for the second ligand molecule). Graphs B and C show the Scatchard and Hill representations respectively. In both panels simulated curves correspond to $k=5\cdot10^5$ M^{-1} and the affinity for the second site is increased (for positive cooperativity) or decreased (negative cooperativity) five times.

Figure 4. Binding to cooperative sites

$$\log\left(\frac{\bar{v}}{n-\bar{v}}\right)_{[L]\to0} = \log\frac{K_1}{n} + \log[L] \qquad (26)$$

At moderate saturation of the macromolecule, the ligand binds to sites where affinity has changed (second and subsequent sites of the macromolecule). As a result, the slope of this region of the Hill curve will change to values higher than one in the case of positive cooperativity, or lower than one for negative cooperativity. Thus, the slope will achieve its maximum experimental value around half of saturation. The maximum theoretical value in this zone will be hypothetically reached in the case of infinite cooperativity, where only the free and fully saturated M species are significantly populated. In this case, it can be demonstrated that this slope can be equal to n. In the real cases where cooperativity is finite, the value will range between $1 < n_H < n$

$$M + nL \xrightarrow{\beta_n} ML_n \qquad\qquad \log\left(\frac{\bar{v}}{n-\bar{v}}\right) = \log K_n + n\cdot\log[L] \qquad (27)$$

- At very high ligand concentrations, reaching saturation of the macromolecule, for almost the totality of macromolecules all sites are occupied by ligand molecules except one of them and the binding to this last site does not influence binding affinity. Therefore, the slope returns again to be equal to 1 (Figure 4) and it can be deduced from the mathematical expression of the Hill equation in the limit of strong binding that

$$\log\left(\frac{\bar{v}}{n-\bar{v}}\right)_{[L]\to\infty} = \log\left(n\cdot K_n\right) + \log[L] \qquad (28)$$

2.6. Physico-chemical description of binding cooperativity

2.6.1. Phenomenological description

This approach does not imply the assumption of any structural model for the ML_i species. For a situation where sites are independent it is assumed that only one microscopic binding constant, k, exists, but there are different macroscopic constants, K_i, for the different equilibriums, according to the scheme shown in Figure 5.

Therefore, the binding polynomial can be generally described as

$$Z = 1 + K_1[L] + K_1K_2[L]^2 + K_1K_2K_3[L]^3 + K_1K_2K_3K_4[L]^4 + \dots \qquad (29)$$

In general, we can also assume that the relationship between macroscopic and microscopic binding constants can be

$$K_i = \frac{Isoforms\ of\ ML_i}{Isoforms\ of\ ML_{i-1}}k \qquad (30)$$

By replacing this equation in the binding polynomial, we obtain the phenomenological expression for this function, and then, by using equation 25 the binding parameter can be estimated. In the case of cooperative sites, this description assumes that the microscopic binding constant changes upon binding, increasing when positive cooperativity happens, and decreasing for negative cooperativity. This approach does not explain the molecular reasons of such a change, as do the following schemes, but it represents an easy way to estimate the equilibrium constants in these cases.

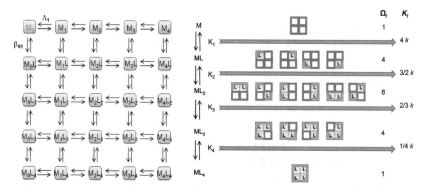

Left side shows the matrix of the model, where columns represent ligand binding equilibriums and rows the conformational changes associated to a hypothetical macromolecule displaying four cooperative binding sites for the ligand. Right side shows the relationship between macroscopic and microscopic constants, where Ω_i represents the number of isoforms of ML_i.

Figure 5. Phenomenological description of binding cooperativity

2.6.2. The Koshland-Nemety-Filmer model

The basics of this model were initially proposed by Pauling to study the cooperative binding of oxygen to haemoglobin [8]. It can explain both, positive and negative cooperative behaviours. More interesting, this model assumes that the different binding sites of the macromolecule are influencing each other through their mutual interconnection by means of a molecular ligature, σ. In Figure 6 we have represented this situation for a macromolecule with four binding sites for the ligand L. It must be considered that when every site is occupied by L, it breaks the ligature with the others, resulting in a modification of their binding affinities. Then, the mathematical expression for the binding polynomial is:

$$Z = \sum_{i=0}^{n} I_i \frac{k^i}{\sigma^{Bi}} [L]^i \qquad (31)$$

where I_i are the number of possibilities of allocating i ligands into the macromolecule, and B_i is the number of broken ligatures for each configuration. The results for the case of a macromolecule with four binding sites are collected into Figure 6 as an example.

This description can be modified as a function of the experimental behaviour of every macromolecule-ligand example. It might be easily developed for the case of different microscopic constants or, even, of different contribution of the ligatures. The main advantage with respect to the phenomenological description is that it can reveal molecular aspects of cooperative phenomena when applied.

2.6.3. The Monod-Wyman-Changeux model

Although this model has been widely used in the literature [4, 9], it can only be used to describe positive cooperativity, which is expressed by assuming that the macromolecule can exists in at least two different conformations, which are under mutual equilibriums, and differ in their affinities for the ligand. Within every conformation, the binding sites behave as if they were equivalent and independent for the binding of L.

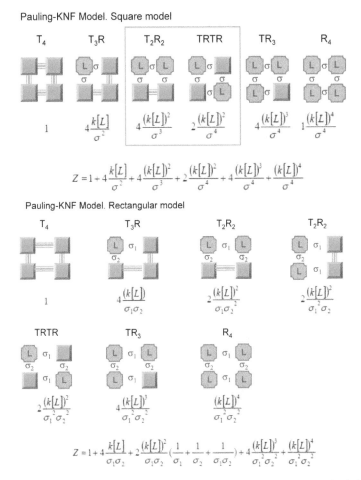

Pauling-KNF Model. Square model

$$Z = 1 + 4\frac{k[L]}{\sigma^2} + 4\frac{(k[L])^2}{\sigma^3} + 2\frac{(k[L])^2}{\sigma^4} + 4\frac{(k[L])^3}{\sigma^4} + \frac{(k[L])^4}{\sigma^4}$$

Pauling-KNF Model. Rectangular model

$$Z = 1 + 4\frac{k[L]}{\sigma_1\sigma_2} + 2\frac{(k[L])^2}{\sigma_1\sigma_2}(\frac{1}{\sigma_1} + \frac{1}{\sigma_2} + \frac{1}{\sigma_1\sigma_2}) + 4\frac{(k[L])^3}{\sigma_1^2\sigma_2^2} + \frac{(k[L])^4}{\sigma_1^2\sigma_2^2}$$

The upper scheme represents the so-called square version of this model for a hypothetical macromolecule displaying fou cooperative binding sites for the ligand and only one kind of ligature. The lower panel shows the rectangular version, with two kinds of ligatures. In both cases we also show the corresponding formulae of the partition function, Z, below each one.

Figure 6. The Koshland-Nemety-Filmer model of binding cooperativity

In Figure 7 a schematic diagram is shown for the case of a macromolecule with four binding sites and two (left side) distinct conformations under equilibrium. In this situation, it is usually assumed that the allosteric equilibrium constant, Λ, is initially big enough to move the equilibrium towards the T-state, considered as the low affinity state. Upon addition of the ligand, the equilibrium moves towards the R-state, of higher affinity than the former. Therefore, this displacement will allow the rest of sites to bind the ligand with higher affinity than the first one. This progressive displacement to the R-state may, thus, explain an increase in affinity (positive cooperative), but not the opposite.

The binding polynomial can be mathematically expressed as

$$Z = \frac{1}{1+\lambda}\left(1 + k_R\left[L\right]\right)^n + \frac{\lambda}{1+\lambda}\left(1 + k_T\left[L\right]\right)^n \tag{32}$$

This model can be easily generalized to more than two conformations of the macromolecule by adding additional terms to this general equation. For example, for the case of three conformations (right side of Figure 7):

$$Z = \frac{1}{1+\lambda+\varpi}\left(1 + k_R\left[L\right]\right)^n + \frac{\lambda}{1+\lambda+\varpi}\left(1 + k_T\left[L\right]\right)^n + \frac{\varpi}{1+\lambda+\varpi}\left(1 + k_S\left[L\right]\right)^n \tag{33}$$

and so on.

3. Notes on ITC performance and general experimental procedures

As was mentioned in the Introduction, ITC is a thermodynamic technique that directly measures the heat released or absorbed in an intermolecular interaction, such as ligand-protein interactions, protein-protein interactions, etc [10]. An ITC experiment consists of a calorimetric titration of a specific volume of one of the reagents, usually the macromolecule, with controlled quantities of the other reagent, usually the ligand, at constant temperature and pressure. Thus, the measured heat during the titration corresponds to the enthalpy of such interactions [11]. This relatively easy experiment allows a complete and precise thermodynamic characterization of the binding event. Subsequently, if the thermal effect is high enough, and the value of the binding constant is moderately good, a single ITC experiment can establish the equilibrium binding constant, K_b, the apparent enthalpy change, ΔH_{app}, and the stoichiometry of the reaction, n. Additionally, if the experiments are made at different temperatures, the change of heat capacity of the process, ΔC_{pb}, can also be measured.

The most common titration calorimeters are adiabatic and are based on the compensation of the thermal effect generated by the addition of the ligand into the sample cell, which is placed in an adiabatic environment [11]. In the left side of Figure 8 we show a schematic representation of one of these instruments. A thermoelectric device measures the temperature difference between the sample and the reference cells (ΔT_{-1}) and also between each cell and the adiabatic jacket (ΔT_{-2}). As long as the reaction is being developed, ΔT_{-1} value decreases to zero with the heating of the sample cell (if the reaction is endothermic) or the reference cell (if exothermic). This heating generates a spike over the baseline of the stationary power, and the integration of this potential required to get ΔT_{-1} to zero in the time to recover the equilibrium is the heat of each injection (right panel in Figure 8).

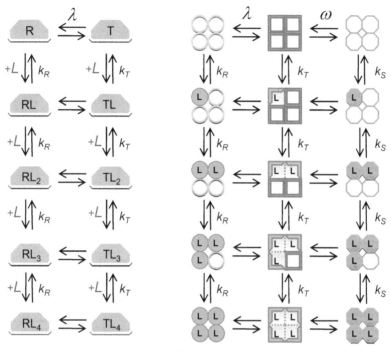

The two schemes show the equilibriums for a hypothetical macromolecule displaying four cooperative binding sites for the ligand and one (left) or two (right) different conformational changes.

Figure 7. The Monod-Wyman-Changeux model of binding cooperativity

As we stated above, a typical ITC experiment consists of a series of injections of determined ligand solution volumes, which can be equal or variable each time, into a macromolecule solution. Such injections have to be separated in between by a time interval, large enough to be sure that the system has reached the equilibrium and all heat absorbed or emitted has been transferred. The titration process is continued until saturation of the macromolecule by the ligand into the cell is reached. In this way, the last additions will not give a significant heat exchange, as shown in Figure 8. The final thermogram is obtained by the individual integration of each peak, setting the integration limits in the baseline that precedes and continues such peak with the equation:

$$Q = \int_{ti}^{tf} W(t)\cdot dt \tag{34}$$

3.1. Procedures for ITC experiments

As we have mentioned previously, this technique allows us to directly evaluate the heat exchange generated upon binding of two molecules. The correct performance of a titration experiment has to consider two main aspects, first, the samples preparation and, second, the proper measurement of reaction heats in the calorimeter. In this Section we are going to describe both experimental aspects.

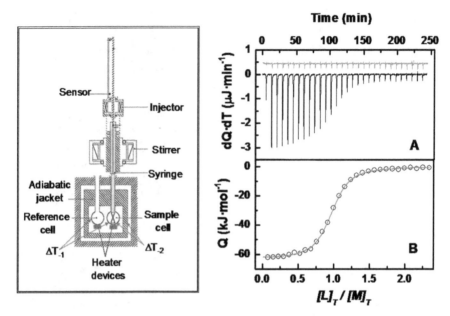

Left side: a schematic diagram of the main components of a titration calorimeter. Right side: a titration calorimetry experiment of a protein with a ligand. In panel A) the titration thermogram is represented as heat per unit of time released after each injection of the ligand into the protein (black), as well as the dilution of ligand into buffer (red). In panel B) the dependence of released heat in each injection *versus* the ratio between total ligand concentration and total protein concentration is represented. Circles represent experimental data and the line corresponds to the best fitting to a model considering n identical and independent sites.

Figure 8. Isothermal titration calorimetry instrumentation

3.1.1. Sample preparation

To carry out ITC binding studies the protein must be of higher purity than 95%, and dialyzed against a selected buffer. Moreover, it is recommended to use the buffer solution of the last dialysis change as the reference solution. The process and precautions to prepare the ligand solution are the same as those described for the protein solution. Nevertheless, when ligands are small as is not possible to dialyze them, the lyophilized (solid) ligand can be directly dissolved with the last dialysis change buffer used for protein sample preparation. As an alternative way, and in order to minimize the differences in the composition of solutions of protein and ligand, lyophilized and solid samples (usually the low molecular weight ligands) can be dissolved in Milli-Q water at a double concentration than that required in the experiment, as well as the protein solution. Accordingly, the dialysis buffer and the protein solution may also contain twice the concentration of buffering salts desired in the experiment (2x buffer). The experimental solutions of protein, ligand and reference buffer (1x buffer) are prepared by adding to the protein and ligand solutions in the necessary amounts (1:1 dilution) of 2x buffer from the last change of dialysis and Milli-Q water respectively, following the subsequent pH correction of both solutions.

The three solutions (buffer, protein and ligand) have to be centrifuged and/or filtered prior to filling the calorimetric cells in order to avoid insoluble particles, it is also recommended to degas them in order to avoid bubbles. The exact protein concentration of the solution has to be determined just before filling the calorimetric sample cell. One of the most accurate and used methods is the spectrophotometric one using protein extinction coefficients (described by Gill & von Hippel [12]).

3.1.2. Modeling and performance of an ITC experiment

Once the protein solution is in the calorimetric cell and the ligand solution in the ITC syringe (where no air bubbles are present), it is very important to wait enough time to be sure that everything is properly thermostated; a way to control such thermal equilibration is controlling the signal of the ITC instrument. Once the signal of the ITC is stable, the experiment is ready to start with the series of ligand injections.

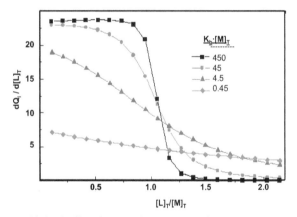

Simulation of heat per added mole of ligand associated to each injection for an ITC experiment with the following experimental parameters: Vc = 1.347 mL; $[M]_T$ = $1.8 \times 10^{-4} M^{-1}$ in the cell; 20 injections of 5µL; $[L]_T$ = 5mM in the syringe; one binding site and four different values of the association constant, 2.5×10^6, 2.5×10^5, 2.5×10^4 and 2.5×10^3 M^{-1}.

Figure 9. Simulations in isothermal titration calorimetry

The limits of a correct determination of the binding thermodynamic parameters using this technique are given by the product of the binding constant, K_b, and the total concentration of the macromolecule, $[M]_T$, being $1 < K_b \cdot [M]_T < 1000$ [11]. Different simulations of a conventional ITC experiment with additions of equal volumes in which four different binding constants have been considered, $2.5 \cdot 10^6$, $2.5 \cdot 10^5$, $2.5 \cdot 10^4$ and $2.5 \cdot 10^3$ M^{-1}, are shown in Figure 9. These values are in the range that includes both high and low affinity (the product

$K_b \cdot [M]_T$ ranges from 450 to 0.45). As it can be observed, when the product $K_b \cdot [M]_T$ is within the appropriate range, the sigmoid curve is obtained, which is needed to perform an analysis with acceptable standard errors. When the product $K_b \cdot [M]_T$ is close to the limits, the isotherm can be optimized with some variations of the experimental design, such as the concentration of total macromolecule in the cell or ligand in the syringe, or also by designing profiles of different injected volumes of ligand. Such profiles usually start with lower ligand volumes for the first injections, which increase progressively in a nonlinear way. Another advantage of using an optimal injection volume profile is the increase of the signal/noise ratio at the end of the thermogram, where the heats of binding are quite small. When the product $K_b \cdot [M]_T$ is over the range, titration experiments by displacement can be performed, in which the target ligand competes for the same binding site with another ligand whose interaction has been previously characterized [13-15].

3.2. Previous treatment of ITC experimental data for thermodynamic analysis

Once the ITC experiment has been performed (black titration in panel A of Figure 8), the thermogram can be integrated to obtain the corresponding heats of each injection. Nevertheless, in order to correct the dilution heat effect of the ligand it is necessary to make a baseline ITC experiment (red titration in panel A of Figure 8), which consist of performing an identical ITC experiment of the ligand binding but with buffer instead of protein into the calorimetric cell. Then, the heats for each injection obtained in this baseline experiment are subtracted to the corresponding ones to the ligand binding experiment. Afterwards, we have to normalize the obtained net heats by the total concentration of ligand in the cell after each injection. The binding isotherm can be obtained by the representation of transferred heat per added mole of ligand ($dQ_i/dL_{T,i}$) versus the molar fraction ($[L]_T/[M]_T$) (panel B in Figure 8).

3.3. Corrections to possible additional heat contributions to the binding experiment

The fitting of the experimental data to the equations explained in the next Section, will allow us to obtain, besides the binding constant or the Gibbs energy, the binding enthalpy. Sometimes this enthalpy change obtained from the fittings of experimental data is the result of additional events occurring during the ligand binding process. So it is important to distinguish between the apparent binding enthalpy, ΔH_{app}, and the real or intrinsic binding enthalpy, ΔH_{int}.

One of the possible events associated to the ligand binding process can be a conformational change of the protein and/or the ligand not associated uniquely to the interaction. A typical example is when the free protein is partially denatured at the experimental temperature, so it is important to check the folding of the protein using other techniques, as circular dichroism or differential scanning calorimetry.

Another possibility is that the ligand binding to the protein can be associated to a change in the pK$_a$ of ionisable groups of the protein or/and the ligand [16-18], in such a way that:

$$\Delta H_{app} = \Delta H_{int} + n_p \cdot \Delta H_{ion} \qquad (35)$$

where n_p is the number of protons accepted or liberated due to the ligand binding to the protein and ΔH_{ion} is the ionization enthalpy of the buffer used in the experiment. As ΔH_{app} depends on the buffer used and the working pH, the easiest way to determine ΔH_{int} is to perform several ITC experiments under the same conditions (mainly at the same pH and ionic strength), but using buffers of different ΔH_{ion}, which permits the determination of the net binding enthalpy from the ordinate of the corresponding linear correlation of ΔH_{app} versus ΔH_{ion}, i.e., the enthalpy value without buffer ionization contributions ΔH_{int}.

4. Thermodynamic analysis of ITC experiments by using different equilibrium models

The analysis of the isotherms is done by the non-linear fitting of the experimental data using different equations, depending on the way the ligand binds to the macromolecule. In this Section we are going to describe, as an example, the four most common models in the literature. The fittings can be done with the appropriate software, as Origin 7.0 (Microcal Software Inc.) or SigmaPlot 2000 (Jandel Co.).

4.1. Ligand binding to one macromolecule with n identical and independent sites

Although the mathematical formulae corresponding to this binding model has been described in Section 2.2.3, we will start defining several functions and parameters. Thus, the binding parameter, $\bar{v}$, being the relationship between the concentration of bound ligand, $[L]_b$, and the total macromolecule concentration, $[M]_T$, can be the one given in equation 18:

$$\bar{v} = \frac{nk[L]}{1 + k[L]} \qquad (36)$$

where k is the microscopic equilibrium constant, which is unique since all binding sites are independent, $[L]$ is the concentration of non-bounded ligand and n is the number of binding sites in the macromolecule.

The heat released or absorbed in any ITC injection, q_i, is related to the binding process as

$$q_i = \Delta H_{app}(\frac{kJ}{molL_b})\Delta(molesL_b) \qquad (37)$$

where ΔH_{app} is the apparent enthalpy change per mole of bound ligand, and $\Delta(molesL_b)$ is the molar amount of ligand bounded in the injection i. If we express the moles of ligand bound in terms of concentrations, the above equation can be written as:

$$q_i = \Delta H_{app} \cdot V_C \cdot ([L]_{b,i} - [L]_{b,i-1}) = \Delta H_{app} \cdot V_C \cdot (\bar{v_i} \cdot [M]_{T,i} - \bar{v}_{i-1} \cdot [M]_{T,i-1}) \qquad (38)$$

where V_c represents the effective volume of the ITC cell and $[M]_T$ is the total concentration of the protein in the cell at injection i.

Furthermore, as known parameters the effective volume of the ITC cell, V_c, the injection volume, V_{inj}, and the ligand concentration in the syringe, $[L]_0$, we can express the concentrations of macromolecule, $[M]_{T,i}$, and ligand, $[L]_{T,i}$, at each injection using the equations

$$[M]_{T,i} = [M]_{T,i-1} \frac{V_C - V_{inj}}{V_C} \qquad\qquad [L]_{T,i} = \frac{(V_C - V_{inj}) \cdot [L]_{T,i-1} + V_{inj} \cdot [L]_0}{V_C} \qquad (39)$$

Thus, the total heat accumulated after N injections could be described as

$$Q = \sum_{i=1}^{N} q_i = V_C \cdot [M]_T \cdot \Delta H_{app} \cdot \bar{v} = V_C \cdot [M]_T \cdot \Delta H_{app} \cdot \frac{n \cdot K \cdot [L]}{1 + K \cdot [L]} \qquad (40)$$

During the ITC experiment the value of the non-bounded ligand concentration, $[L]$, is an unknown variable and for this reason, it is operationally required to estimate it from the experimental variables $[L]_T$ and Q as

$$[L] = [L]_T - [L]_b = [L]_T - \frac{Q}{V_C \cdot \Delta H_{app}} \qquad (41)$$

Substituting the above equation in equation 40 we obtain a quadratic equation with Q as unknown variable, whose solution is

$$Q = \frac{Vc \cdot \Delta H_{app}}{2 \cdot k} \left[1 + k \cdot [L]_T + n \cdot k \cdot [M]_T - \sqrt{(1 + k \cdot [L]_T + n \cdot k \cdot [M]_T)^2 - 4 \cdot n \cdot k^2 [M]_T \cdot [L]_T} \right] \qquad (42)$$

Finally, deriving this expression with respect to $[L]_T$ we obtain an expression for the heat per mole of ligand added in each injection

$$\frac{1}{V_c} \cdot \frac{dQ}{d[L]_T} \approx \frac{1}{V_c} \cdot \frac{\Delta Q}{\Delta[L]_T} = \frac{\Delta H_{app}}{2} \left[1 - \frac{1 + [L]_T - n \cdot k \cdot [M]_T}{\sqrt{(1 + k \cdot [L]_T + n \cdot k \cdot [M]_T)^2 - 4 \cdot n \cdot k^2 \cdot [M]_T \cdot [L]_T}} \right] \qquad (43)$$

According to these equations, there are two possible ways to analyze the experimental heats from an ITC experiment: one by using equation 43 which considers the heat per mole of added ligand associated with each injection; the second by using equation 42 and considering the total heat accumulated from the beginning to each injection of the ITC experiment. The first approach has the advantage of avoiding experimental errors, since in such analysis is possible to eliminate individual experimental points from the curve (Figure 8), while the second approach imply the sum of all the heats of each injection which accumulates errors.

4.2. Ligand binding to one macromolecule with m different and independent classes of sites

In this model, each binding site is defined as an independent site, with different affinity to the other binding sites. The expression "different sites" implies a microscopic equilibrium constant for each binding site, and the term "independent" site means that the binding affinity does not change with the binding of any other ligand to the other sites of the macromolecule. The mathematical formulae that we describe here correspond to a macromolecule with only two different classes of sites (m = 2) with n_1 and n_2 sites for each type, as represented in the following scheme:

$$M + \left(n_1 + n_2\right)L \xleftrightarrow{K1 \cdot K2} ML_{n1+n2} \tag{44}$$

The binding parameter, defined as the ratio of the concentration of ligand bound at any of the two classes of sites, $[L]_{b,i}$, and the total concentration of macromolecule, $[M]_T$, can be expressed now as

$$\bar{v} = \sum_{i=1}^{m=2} \bar{v}_i = \frac{[L]_{b,i}}{[M]_T} = \sum_{i=1}^{m=2} \frac{n_i \cdot K_i \cdot [L]}{1 + K_i \cdot [L]} \tag{45}$$

Thus, the heat released or absorbed in any injection, q_j, would be

$$q_j = \sum_{i=1}^{m=2} \Delta H_{app,i}\left(\frac{kJ}{molL_{b,i}}\right) \cdot \Delta\left(molesL_{b,i}\right) \tag{46}$$

where $\Delta H_{app,i}$ is the apparent enthalpy change per mole of ligand bound to any of the two classes of sites. If we express the moles of ligand bound in terms of concentrations, then the above equation can be re-formulated as:

$$q_j = \sum_{i=1}^{m=2} \Delta H_{app,i} \cdot V_c \cdot ([L]_{b,i,j} - [L]_{b,i,j-1}) = \sum_{i=1}^{m=2} \Delta H_{app,i} \cdot V_c \cdot (\bar{v}_j \cdot [M]_{T,j} - \bar{v}_{i-1} \cdot [M]_{T,j-1}) \tag{47}$$

where V_c represents the effective volume of the ITC cell and $[M]_{T,j}$ is the concentration of protein in the cell after injection j.

Thus, if we substitute equations 39 in the above expression, we obtain the following:

$$Q = \sum_{j=1}^{N} q_j = V_c \cdot [M]_T \cdot \sum_{i=1}^{m=2} \Delta H_{app,i} \cdot \bar{v}_{i,N} = V_c \cdot [M]_T \cdot \sum_{i=1}^{m=2} \Delta H_{app,i} \frac{n_i \cdot k_i \cdot [L]}{1 + k_i \cdot [L]} \tag{48}$$

Solving the summation for two classes of sites, m=2, the expression of the total heat accumulated in N injections can be re-written as

$$Q = V_c \cdot [M]_T \left[\Delta H_{app,1} \frac{n_1 \cdot k_1 \cdot [L]}{1 + k_1 \cdot [L]} + \Delta H_{app,2} \frac{n_2 \cdot k_2 \cdot [L]}{1 + k_2 \cdot [L]}\right] \tag{49}$$

Since the value of [L] is unknown, we should express it in terms of total bound ligand ($[L]_{b,T} = [\overline{v_1 + v_2}] \cdot [M]_T$), as we show in the following equation

$$[L] = [L]_T - [L]_{b,T} = [L]_T - [M]_T \cdot \left[\frac{n_1 \cdot k_1 \cdot [L]}{1 + k_1 \cdot [L]} + \frac{n_2 \cdot k_2 \cdot [L]}{1 + k_2 \cdot [L]} \right] \tag{50}$$

Substituting the previous expression in the equation 49 and re-organizing it, we obtain the following cubic equation:

$$[L]^3 + a_2 [L]^2 + a_1 [L] + a_0 = 0 \tag{51}$$

where the coefficients a_0, a_1 and a_2 are defined as

$$a_0 = -\frac{[L_T]}{k_1 k_2}$$

$$a_1 = \left(\frac{n_1}{k_2} + \frac{n_2}{k_1} \right) [M_T] - \left(\frac{1}{k_1} + \frac{1}{k_2} \right) [L_T] + \frac{1}{k_1 k_2} \tag{52}$$

$$a_2 = \frac{1}{k_1} + \frac{1}{k_2} + (n_1 + n_2) [M_T] - [L_T]$$

The only valid solution to the cubic equation 51 can be simply written by just grouping the coefficients a_0, a_1 and a_2 in three new coefficients A, B and C as

$$[L] = \sqrt[3]{A + \sqrt{A^2 + B^3}} + \sqrt[3]{A - \sqrt{A^2 + B^3}} + C \tag{53}$$

where A, B and C are

$$A = \frac{-a_2^3}{27} + \frac{a_1 a_2}{6} - \frac{a_0}{2} \qquad B = \left(\frac{a_1}{3} - \frac{a_2^2}{9} \right) \qquad C = -\frac{a_2}{3} \tag{54}$$

The solution of the cubic equation (using equations 52 to 54) allows us to calculate the non-bounded ligand concentration for a given number of injections. Substituting in equation 49, we can determine the heat associated to each injection using the following expression:

$$dQ \approx \Delta Q(j) = Q_T(j) - Q_T(j-1) + \frac{V_{in}}{V_C} \left(\frac{Q_T(j) + Q_T(j-1)}{2} \right) \tag{55}$$

4.3. Ligand binding by the displacement of another ligand in the single binding site of a macromolecule

To formulate this binding model of displacement, we assume two ligands, A and B, which can bind to the same binding site of a protein, M, with different affinity constants. Then, we can describe the equilibrium binding for each ligand as:

$$M + A \xleftarrow{K_A} MA \qquad K_A = \frac{[MA]}{[M]\cdot[A]} \qquad K_B = \frac{[MB]}{[M]\cdot[B]} \qquad (56)$$

$$M + B \xleftarrow{K_B} MB$$

Because the binding of two ligands takes place in the same binding site of the macromolecule, and having the ligand A tighter affinity than ligand B, $K_A \gg K_B$, we should consider the following scheme

$$M + B \xleftarrow{K_B} MB \qquad (57)$$

$$MB + A \xleftarrow{K_A} MA + B$$

According to the schemes 56 and 57 we can express the initial concentration of the ligands A and B as

$$[A]_0 = [MA] + [A] \qquad\qquad [B]_0 = [MB] + [B] \qquad (58)$$

Substituting these expressions in equations 56, the association constants can be re-written as

$$[MA] = \frac{[M]\cdot[A]_0}{1/K_A + [M]} \qquad\qquad [MB] = \frac{[M]\cdot[B]_0}{1/K_B + [M]} \qquad (59)$$

It is important to consider that in the case we propose for this binding model, it is usual that the binding of the high-affinity ligand A has been previously analyzed in a simple titration experiment. The displacement titration experiment will allow us to analyze the interaction of the low affinity ligand B, which cannot be determined by direct titration experiments. Thus, initially, the macromolecule M is bounded to ligand B forming the MB complex and during the titration of the ligand A we will shift partially the MB complex formation to the formation of the MA complex.

For the mathematical formulae of this model, we first define the molar fractions of all species containing the macromolecule. Such fractions are

$$x_M = [M]/[M]_T \qquad x_{MA} = [MA]/[M]_T \qquad x_{MB} = [MB]/[M]_T \qquad (60)$$

where $[M]_T$ is the total concentration of macromolecule.

If we also write the molar ratios between the initial amounts of A and B relative to the total macromolecule concentration as

$$r_A = [A]_0/[M]_T \qquad r_B = [B]_0/[M]_T \qquad (61)$$

we can write the concentrations of A and B bounded ligands during the interaction as products of the association constants

$$c_A = K_A \cdot [M]_T \qquad c_B = K_B \cdot [M]_T \qquad (62)$$

Then, substituting these expressions in equations 59, we obtain the following equations for all the species that form the macromolecule, expressed in terms of the molar ratios of macromolecule:

$$x_M + x_{MA} + x_{MB} = 1 \qquad x_{MA} = \frac{r_A \cdot x_M}{1/c_A + x_M} \qquad x_{MB} = \frac{r_B \cdot x_M}{1/c_B + x_M} \qquad (63)$$

Substituting the above equations of x_{MA} and x_{MB} into the first one and rearranging, we obtain the following cubic equation:

$$x_M^3 + a \cdot x_M^2 + b \cdot x_M + c = 0 \qquad (64)$$

in which we have defined the a, b and c coefficients as:

$$a = \frac{1}{c_A} + \frac{1}{c_B} + r_A + r_B - 1 \qquad b = \frac{r_A - 1}{c_A} + \frac{r_B - 1}{c_B} + \frac{1}{c_A \cdot c_B} \qquad c = -\frac{1}{c_A \cdot c_B} \qquad (65)$$

Solving such cubic equation, we obtain the following real solution for the molar fraction of macromolecule:

$$x_M = \frac{2 \cdot (\sqrt{a^2 - 3b}) \cdot \cos(\phi / 3) - a}{3} \qquad (66)$$

in which the coefficient ϕ can be written as:

$$\phi = \arccos \frac{-2 \cdot a^3 - 9 \cdot a \cdot b - 27 \cdot c}{2 \cdot (\sqrt{a^2 - 3 \cdot b})^3} \qquad (67)$$

Once the molar fraction of free macromolecule, x_M, is determined, we can also know the molar fractions of the other species in which the macromolecule is present (x_{MA} and x_{MB}) by solving the equations 63.

The heat released or absorbed after each injection is proportional to the changes in concentration of MA and MB, [MA] and [MB], and their molar enthalpies of binding. Therefore, the heat after each injection can be written according to the following expression

$$\Delta Q = V_c \cdot (\Delta H_A \cdot \Delta [MA] + \Delta H_B \cdot \Delta [MB]) = V_c \cdot [M]_T \cdot (\Delta H_A \cdot \Delta x_{MA} + \Delta H_B \cdot \Delta x_{MB}) \qquad (68)$$

where V_c is the effective volume of the ITC cell. To correct the concentrations of macromolecule and ligand, we define the following infinitesimal change for their concentrations

$$d[X] = -\frac{dV_i}{V_0}[X] \qquad (69)$$

where [X] represents the concentration of any species. Integrating the above expression between the limits from $[X]_i$ to $[X]_{i-1}$ and from zero to V_i. The resulting equation is

$$[X]_i = [X]_{i-1} \exp\left(-\frac{V_i}{V_0}\right) = [X]_{i-1} f_i \qquad (70)$$

where f_i is the dilution factor that allow us to define the molar ratios after each injection as

$$\left[A\right]_i = \left[A\right]_{i-1}(1 - f_i) \qquad \left[M\right]_i = f_i\left[M\right]_{i-1} \qquad \left[B\right]_i = f_i\left[B\right]_{i-1} \qquad (71)$$

Consequently, the heat absorbed or released after each injection can be expressed as

$$dQ \approx \Delta Q = V_C\left[M\right]_T \cdot \left\{\Delta H_A \cdot (\Delta x_{MA,i} - f_i \cdot x_{MA,i-1}) + \Delta H_B \cdot (\Delta x_{MB,i} - f_i \cdot x_{MB,i-1})\right\} \qquad (72)$$

4.4. Ligand binding to a macromolecule with two dependent (cooperative) binding sites

The scheme described in Figure 2 for a macromolecule with two binding sites was used in Section 2.2.2 to describe the correlation between macroscopic and microscopic equilibrium constants and the binding parameter was obtained for the case of independent sites, characterized by the same microscopic constant. Nevertheless, when cooperativity exists among sites, this simple assumption cannot be considered, and more than one value for such constants must be taken into account. Thus, positive cooperativity would be revealed when $k_3 > k_1$ and $k_4 > k_2$, and the opposite is true for negative cooperativity. In any case, it can be easily deduced that $k_3 \cdot k_1 = k_4 \cdot k_2$.

The most simplified version of this model could be attained for equivalent binding sites, that is, when microscopic binding to the first site is identical, independently if the ligand binds to either site 1 or 2. The occupancy of the first site will drive to the modification of the affinity of the second one, in a similar way in both branches of the scheme. Therefore, $k_1 = k_2$ and $k_3 = k_4$. In parallel, the enthalpy changes associated can be also grouped in only two different values, the changes for the formation of ML and ML$_2$ species respectively. From here, any other version should be much more complicated from a mathematical point of view. Overcoming these calculation matters, the solution can be achieved in a similar way to the described here for the simplest version, which is also inspired in the models described previously.

The molar fractions of all species containing the macromolecule are in this case

$$x_{ML} = \frac{\left[ML\right]}{\left[M\right] + 2\left[ML\right] + \left[ML_2\right]} = \frac{k_A\left[L\right]}{1 + 2k_A\left[L\right] + k_A k_B\left[L\right]^2}$$

$$x_{ML2} = \frac{\left[ML_2\right]}{\left[M\right] + 2\left[ML\right] + \left[ML_2\right]} = \frac{k_A k_B\left[L\right]^2}{1 + 2k_A\left[L\right] + k_A k_B\left[L\right]^2} \qquad (73)$$

where we have assumed that $k_1 = k_2 = k_A$ and $k_3 = k_4 = k_B$.

Taking into account the equation 40, the total heat accumulated after N injections could be described as

$$Q = \sum_{i\_1}^{N} q_i = V_C \cdot [M]_T \cdot \Delta H_{app} \cdot \overline{v} = V_C \cdot [M]_T \cdot \left\{x_{ML}\left(\Delta H_A + \Delta H_B\right) + x_{ML2}\left(\Delta H_A + \Delta H_B\right)\right\} \qquad (74)$$

where ΔH_A and ΔH_B are the enthalpy changes associated to equilibriums characterized by k_A and k_B respectively. Since the value of [L] is unknown, we should express it in terms of total ligand ($[L]_T = [L] + 2[ML] + 2[ML_2]$), as we show in the following equation

$$[L] = [L]_T - 2 \cdot x_{ML}[M]_T - 2 \cdot x_{ML2}[M]_T \tag{75}$$

This can be expressed as the following third-order equation:

$$[L]^3 + a_2[L]^2 + a_1[L] + a_0 = 0 \tag{76}$$

where the coefficients a_0, a_1 and a_2 are defined as

$$a_0 = -\frac{[L]_T}{k_A k_B} \qquad a_1 = \frac{2}{k_B}[M]_T - \frac{2}{k_B}[L]_T + \frac{1}{k_A k_B} \qquad a_2 = \frac{2}{k_B} + 2[M]_T - [L]_T \tag{77}$$

The only valid solution to the cubic equation above can be simply written by grouping the coefficients a_0, a_1 and a_2 in three new coefficients A, B and C as follows:

$$[L] = \sqrt[3]{A + \sqrt{A^2 + B^3}} + \sqrt[3]{A - \sqrt{A^2 + B^3}} + C \tag{78}$$

where A, B and C are

$$A = \frac{-a_2^3}{27} + \frac{a_1 a_2}{6} - \frac{a_0}{2} \qquad B = \left(\frac{a_1}{3} - \frac{a_2^2}{9}\right) \qquad C = -\frac{a_2}{3} \tag{79}$$

The solution of the cubic equation (using equations 77 to 79) allows us to calculate the non bounded ligand concentration for a given number of injections. Substituting in equation 74, we can determine the heat associated to each injection using the following expression:

$$dQ \approx \Delta Q(j) = Q_T(j) - Q_T(j-1) + \frac{V_{in}}{V_C}\left(\frac{Q_T(j) + Q_T(j-1)}{2}\right) \tag{80}$$

4.5. Guidelines for the development of ITC equilibrium models

Following the reasoning given in this Chapter, it is easy to discern the basic rules to build any ITC model. The main point would be to collect any experimental and structural evidence (number of sites in the macromolecule M for ligand L, their dependent or independent character, etc) to develop the basic interaction scheme. This scheme will drive to the construction of the binding polynomial, Z, which derives into the binding parameter, $\bar{v}$, as it has been described in detail into Section 2.

The examples given into Section 4 for the most common ITC models used reveal that, once the model is described in terms of the binding parameter or by the molar fractions of all species containing the macromolecule, two basic points have to be solved from a mathematical point of view. That is, the total heat accumulated after N injections as a function of the binding parameter (equation 40) and the concentration of free ligand as a function of the well-known total concentrations of macromolecule and ligand. Both solutions can be replaced into an equation such as 80 to determine the heat associated to each injection.

These heats divided by the number of moles of ligand added represent the dependent variable of the curve fitting analysis (dQ_i/dL_i), where the ratio between the total concentrations of ligand and macromolecule $[L]_T/[M]_T$ represents the independent variable of the fitting function. The results of this non-linear regression analysis provide the values of the equilibrium constants and the respective enthalpy changes involved in such equilibriums, according to the proposed model. An example is illustrated in Figure 8.

Author details

Jose C. Martinez, Javier Murciano-Calles, Eva S. Cobos, Manuel Iglesias-Bexiga,
Irene Luque and Javier Ruiz-Sanz
Department of Physical Chemistry and Institute of Biotechnology, Faculty of Sciences,
University of Granada, Granada, Spain

Acknowledgement

This work was financed by grant CVI-5915 from the Andalucian Regional Goverment (Spain), grant BIO2009-13261-C02-01 from the Spanish Ministry of Science and Technology, FEDER and Plan E.

5. References

[1] Langerman N & Biltonen RL (1979) Microcalorimeters for biological chemistry: applications, instrumentation and experimental design. Methods Enzymol 61: 261-286.

[2] Biltonen RL & Langerman N (1979) Microcalorimetry for biological chemistry: experimental design, data analysis, and interpretation. Methods Enzymol 61: 287-318.

[3] Wyman J (1965) Binding Potential a Neglected Linkage Concept. Journal of Molecular Biology 11: 631-&.

[4] Hess VL & Szabo A (1979) Ligand-Binding to Macromolecules - Allosteric and Sequential Models of Cooperativity. Journal of Chemical Education 56: 289-293.

[5] Szabo A & Karplus M (1972) Mathematical-Model for Structure-Function Relations in Hemoglobin. Journal of Molecular Biology 72: 163-&.

[6] Szabo A & Karplus M (1975) Analysis of cooperativity in hemoglobin. Valency hybrids, oxidation, and methemoglobin replacement reactions. Biochemistry 14: 931-940.

[7] Szabo A & Karplus M (1976) Analysis of Interaction of Organic-Phosphates with Hemoglobin. Biochemistry 15: 2869-2877.

[8] Pauling L (1935) The Oxygen Equilibrium of Hemoglobin and Its Structural Interpretation. Proc Natl Acad Sci U S A 21: 186-191.

[9] Cuadri-Tome C, Baron C, Jara-Perez V, Parody-Morreale A, Martinez JC & Camara-Artigas A (2006) Kinetic analysis and modelling of the allosteric behaviour of liver and muscle glycogen phosphorylases. J Mol Recognit 19: 451-457, doi: 10.1002/jmr.772.

[10] Ladbury JE & Chowdhry BZ (1996) Sensing the heat: the application of isothermal titration calorimetry to thermodynamic studies of biomolecular interactions. Chem Biol 3: 791-801.

[11] Wiseman T, Williston S, Brandts JF & Lin LN (1989) Rapid measurement of binding constants and heats of binding using a new titration calorimeter. Anal Biochem 179: 131-137.

[12] Gill SC & von Hippel PH (1989) Calculation of protein extinction coefficients from amino acid sequence data. Anal Biochem 182: 319-326.

[13] Sigurskjold BW (2000) Exact analysis of competition ligand binding by displacement isothermal titration calorimetry. Anal Biochem 277: 260-266, doi: 10.1006/abio.1999.4402S0003-2697(99)94402-0 [pii].

[14] Velazquez-Campoy A & Freire E (2006) Isothermal titration calorimetry to determine association constants for high-affinity ligands. Nat Protoc 1: 186-191, doi: nprot.2006.28 [pii]10.1038/nprot.2006.28.

[15] Zhang YL & Zhang ZY (1998) Low-affinity binding determined by titration calorimetry using a high-affinity coupling ligand: a thermodynamic study of ligand binding to protein tyrosine phosphatase 1B. Anal Biochem 261: 139-148, doi: S0003-2697(98)92738-5 [pii]10.1006/abio.1998.2738.

[16] Baker BM & Murphy KP (1996) Evaluation of linked protonation effects in protein binding reactions using isothermal titration calorimetry. Biophysical Journal 71: 2049-2055.

[17] Mason AC & Jensen JH (2008) Protein-protein binding is often associated with changes in protonation state. Proteins-Structure Function and Bioinformatics 71: 81-91, doi: Doi 10.1002/Prot.21657.

[18] Velazquez-Campoy A, Luque I, Todd MJ, Milutinovich M, Kiso Y & Freire E (2000) Thermodynamic dissection of the binding energetics of KNI-272, a potent HIV-1 protease inhibitor. Protein Sci 9: 1801-1809, doi: 10.1110/ps.9.9.1801.

Chapter 5

Insights into the Relative DNA Binding Affinity and Preferred Binding Mode of Homologous Compounds Using Isothermal Titration Calorimetry (ITC)

Ruel E. McKnight

Additional information is available at the end of the chapter

1. Introduction

1.1. Drug-DNA interactions

Many biologically significant compounds have been known, for several decades now, to bind non-covalently to nucleic acids.[1-7] Ever since the discovery of the structure of DNA in the 1950s, DNA has been a target for many therapeutic compounds. Several of these compounds have been found to bind to DNA while interfering with the activity of many vital enzymes and protein factors involved in DNA metabolism. Others cleave DNA or cause DNA cross-linking (for example, cisplatin) interfering with cell division and leading to apoptosis. As a result, several DNA-binding compounds have been identified as therapeutic agents in especially the anti-cancer and anti-pathogenic classes. Some of the most notable members of these classes include the Streptomyces derived anthracyclins e.g., daunomycin (daunorubicin) and doxorubicin, have been used for decades, initially as antibiotics, then mainly as antitumor agents.[8] Other known DNA binding agent include mitoxantrone, which has been particularly useful in the treatment of breast cancers, the glycopeptide antibiotic bleomycin which has been used in the treatment of Hodgkin's lymphoma and testicular cancer, amsacrine, bisantrene and various porphyrin derivatives. Even though many of these compounds have exhibited therapeutic potency, there still exist the accompanying unwanted side-effects, due mainly to the lack of selectivity and DNA targeting. Now, even after decades of studies of drug-DNA interactions, the existence of deleterious side-effects remains a huge area of concern and presents the main barrier for progress within the field. So, the question of whether a certain molecule will bind to a specific DNA sequence is currently being probed by several research groups. If we are to approach the problem from a fundamental level, such efforts must rely heavily on a fundamental understanding of the predominant contributions to drug-DNA interactions. Although ligand-DNA interactions have been studied, so far there have only been a handful of studies that have probed the factors that govern DNA-binding using homologous series of compounds. This information is especially relevant to the rationale design of novel therapeutics with improved efficacy and specificity. The proposed chapter is designed to yield an understanding of how various features of small molecules govern their binding to DNA and will provide insights into ligand-DNA interactions by studying binding trends within homologous series of compounds. Several studies have suggested that some DNA binding molecules exhibit more than one binding mode while binding in a sequence specific manner. In fact, some researchers have proposed that the therapeutic efficiency of these drugs may be linked to their ability to exhibit mixed binding modes.[9,10] These modes primarily involve intercalation, where planar aromatic molecules slide between adjacent DNA base pairs resulting in significant perturbation of the DNA, and/or minor-groove binding, where molecules with the requisite flexibility and isohelicity with the DNA minor groove are able to fit into the DNA groove, usually with no significant change in the structure of the DNA.

For many years now, microcalorimetry has been utilized to decipher the complete thermodynamic profiles for a number of drug-DNA complexes.[11] Isothermal titration calorimetry (ITC) has been successfully used to parse the thermodynamics of the interactions between drug molecules and DNA.[2,3,11] ITC is regarded as the "gold standard" approach for the determination of binding affinity data in biomolecular interactions. ITC has been used to determine the comprehensive thermodynamic profile of these interactions, by determining enthalpy change (ΔH) directly (usually in the presence of an excess of the macromolecular binding sites), while determining equilibrium binding constant (K), and number of binding sites (n) by model-fitting routines. Free energy change (ΔG) and ultimately entropy change (ΔS) are determined from the known thermodynamic relationships ($\Delta G = -RT\ln K$) and ($\Delta G = \Delta H - T\Delta S$), respectively. Furthermore, heat capacity change (ΔCp) may be determined from ITC measurements of ΔH over a range of different temperatures ($\Delta Cp = d\Delta H/dT$).[11]

In this chapter, we show how isothermal titration calorimetry can be successfully utilized to determine relative DNA binding efficacy, as well as the preferred DNA binding mode for a selection of homologous series of compounds. By comparing the DNA binding characteristics of homologous compounds under identical conditions, we can make robust conclusions as to the most important driving force governing the interaction of ligands to DNA. The chapter will describe two classes of homologous compounds; the naphthalene diimides and chalcogenoxanthyliums. However, the chapter will mainly focus on the naphthalene diimide series. The NDI scaffold has been used by several researchers to design therapeutically significant candidates [12-20] and are used in our studies as model systems to gain additional insight into the binding of "threading" intercalators to DNA. These symmetrical molecules have two substituents on either side of the intercalating moiety, thus necessitating the threading through or involvement of the side chain during binding

(**Figure1**).[13,14,17-19] In this geometry, one side chain occupies the minor groove, while the other lies in the major groove. According to some researchers, a threading intercalator has a number of potential advantages since the contact with both DNA grooves provide additional potential sites for recognition and targeting.[12-20] The NDI scaffold has provided a versatile template for the design of many promising derivatives.[12-20]

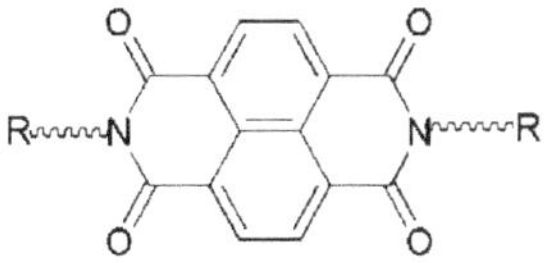

(R = ethyl- or propyl-amino side chain)

Figure 1. General structure of the naphthalene diimides in this study.

Some NDI derivatives have also been found to selectively bind non-standard structural forms of DNA such as triplexes and G-quadruplexes, which are normally transient and unstable.[21-25] Stabilization of DNA triplexes formed when oligonucleotides (normally referred to as triplex formation oligonucleotide or TFO) bind to DNA duplexes, have been explored in anti-gene therapeutics where expression of deleterious DNA sequences are suppressed by the binding and stabilization of complimentary TFO sequences.[15,21] Formation of transient G-quadruplexes in G-rich sequences have been found to be prominent in telomeres, G-rich ends on chromosomes that protects indispensable genes from being depleted, as well as preventing unwanted chromosomal fusions.[23-25] As a result, some compounds (e.g., certain NDI derivatives) can bind to and stabilize these telomeric G-quadruplexes can block access to these sequences by telomerase enzymes, which are responsible for extending and protecting telomeres and have been found to be over-expressed in 80% of cancers cells.[24,25] G-quadruplexes have also been found to be prominent in promoter regions, especially in the promoters of oncogenes such as the *c-myc and Ras genes, were*, found to be directly linked to the formation of certain cancers.[24,25] Stabilization of these G-quadruplexes in oncogene promoter regions can block access by RNA polymerase, and ultimately blocking expression of these deleterious genes. It is therefore important that we continue to probe ligands systems in order to increase our understanding of the driving force behind ligand–DNA interactions, and to use this knowledge to control their preferred binding mode and sequence.

The NDI compounds were synthesized as previously described.[26] As mentioned above, the NDI scaffold has been used by several groups to design biologically significant compounds.[12-20] In the current series, the quaternary amino group in each side chains is close enough (ethyl- and propyl-amino linker) to the naphthalene core group to allow electrostatic contact with the DNA. Therefore, the cationic quaternary amino groups are close to the DNA when the core ring system intercalates between DNA base pairs. As a result, there is a greater probability for electrostatic interaction with the phosphates in the DNA backbone. The NDI molecules of this study have two substituents on either side of the central naphthalene moiety and differ mainly in substituent size and hydrophobicity. That means, each compound should adopt a threading molecular geometry when bound to DNA via intercalation. Threading NDI compounds analogous to the ones in this study have been under investigation for several years as potential therapeutic (especially anticancer) compounds that bind to DNA with improved sequence-selectivity due to their interactions with both DNA grooves.

1.2. Chalcogenoxanthyliums

Although stored blood used during surgery and in blood transfusion is generally safe due to improved screening procedures, there is still a chance (a slight risk) that pathogens within the stored blood may be transmitted from donor to recipient.[27,28] This can occur if the blood was collected from an infected individual before there were detectable levels of the causative pathogen. As a result, there remains a need to develop protocols in which to reduce the risk of pathogen transmission, if only in a precautionary or preventative role.

Photodynamic therapy (PDT) is one approach that has been considered as a viable means in which to purge stored blood samples of deleterious pathogens.[27-32] In PDT, light is used along with endogenous oxygen and an appropriate photosensitizer (a molecule that has the ability to absorb light energy, i.e., photoexcitation, and transfer this energy to another chemical entity inducing a change) to treat or reduce an affliction. Photosensitizers are effective mainly because they are able to absorb appropriate light energy and produce excited triplet states at which time they can transfer energy to ground state oxygen (which is also triplet state) via intersystem crossing producing very toxic singlet oxygen species. PDT has been used for years in the treatment of certain cancers and lesions, as well as age-related macular degeneration. Photofrin, a hematoporhyrin belonging to the porphyrin class of compounds, is probably the most well-known and has been used for many years to treat bladder cancers. Other photosensitizers include those in the clorin class (e.g., photochlor), as well as dyes such as phthalocyanine.

PDT can be applied in pathogen reduction, especially in the removal of microbial material from blood products. In this application, PDT is normally referred to as photodynamic antimicrobial chemotherapy (PACT). Compounds containing the xanthylium core (rhodamines and rosamines), are among some of the most highly touted class of compounds being considered for PACT and have been explored by Wagner, Detty and coworkers.[27,31,32] These compounds have been found to selectively accumulate in cancer cells and mitochondria, and have also been considered as p-glycoprotein inhibitors and mitochondrial stains.[33,34] However, the parent rhodamines and rosamines have been mostly ineffective due to short-lived and low yield of triplet excited state upon photo-excitation. Detty and coworkers have synthesized a group of related chalcogenoxanthyliums (**Figure 2**) that are based on the parent compounds.[33,34]

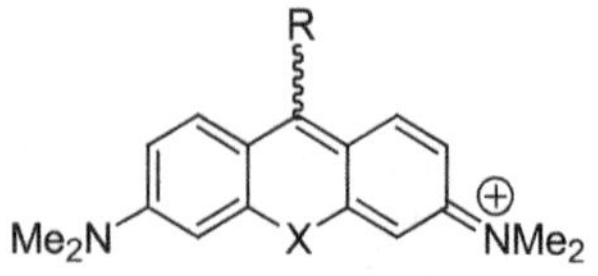

(X = chalchogen, R = 9-aryl substituent)

Figure 2. General structure for the chalcogenoxanthylium derivatives.

These chalcogenoxanthylium derivatives represent an improvement over the parent rhodamine and rosamine since the inclusion of the heavier chalcogen (e.g.,S and Se instead of O) provides the known "heavy atom effect" which increases the production of long-lived excited triplet states.[33] Furthermore, the substituents (for example, a 2-thienyl instead of a phenyl) in the 9-position can be "tuned" such that they absorb light at wavelength that avoids hemoglobin attenuation. [33-35]

To date, PACT has been mostly unsuccessful due largely to 1) low efficacy against pathogens, and 2) unwanted background hemolysis of red blood cells.[32] Both these shortcomings are mostly due to the non-specific actions of the photosensitizers when exposed to the requisite light. To circumvent these problems, photosensitizers that are able to target the pathogenic DNA relative to the red blood cells are currently being explored.[32,35] One approach to target these pathogens in the presence of red blood cells is to use photosensitizers that bind strongly to the pathogenic DNA, since mature red blood cells do not contain organelles or genomic nucleic acids.[32,35] The chalcogenoxanthylium derivatives are advantageous to use since their substituents can be tuned such that 1) they absorb light in a spectral region where light attenuation by hemoglobin absorption is avoided, 2) increased yield of singlet excited state that are responsible for destruction of pathogens, and 3) their planarity and hydrophobicity can be altered to monitor the effects on their interaction with DNA. Thus, offering greater opportunity to potentially reduce the incidence of background hemolysis. The DNA binding efficacy and preferred mode of binding of a series of chalcogenoxanthylium dyes were investigated by isothermal titration calorimetry (ITC).[35]

1.3. Preference for AT-rich vs GC-rich DNA

In an effort to decipher the preferred DNA binding mode for compounds in this study, a preference for an AT- vs GC-rich sequence will be determined. In order to differentiate preferences for intercalation and/or groove binding, the binding of the compounds of this study to [poly(dAdT)]$_2$ and [poly(dGdC)]$_2$ were examined by ITC. Figure 3 shows the structure of [poly(dAdT)]$_2$ and [poly(dGdC)]$_2$ used in this study. It has long been established that known groove binding compounds (e.g., distamycin, berenil, and DAPI) show a strong preference (an order of magnitude or greater) for binding to [poly(dAdT)]$_2$ relative to [poly(dGdC)]$_2$.[6] The lower affinity for GC-rich sequences shown by groove binders is largely due to their restricted access to the minor groove of GC sequences caused by the protruding 2-NH$_2$ group of guanine. Intercalators are only expected to be affected by this if a substituent is placed into the minor groove during formation of the intercalation complex. It is however expected that compounds that exhibit mixed binding mode (i.e., intercalation and groove binding) will exhibit less (<10) of a preference for the AT sequence.[28,35,36]

5'-ATATATATATATATATATAT-3'   5'-GCGCGCGCGCGCGCGCGCGC-3'
3'-TATATATATATATATATATA-5'   3'-CGCGCGCGCGCGCGCGCGCG-5'

Figure 3. AT-rich and GC-rich DNA sequences used in this study.

In this chapter, calorimetric data of naphthalene diimide derivatives binding to both calf thymus DNA (ctDNA), as well as AT- and GC-rich DNA sequences will be described. The binding characteristics of selected chalcogenoxanthylium derivatives will also be compared. In an effort to gain insight into the involvement of a minor groove vs. intercalative binding mode, the binding of the compounds to [poly(dAdT)]$_2$ and [poly(dCdG)]$_2$ sequences (using ITC) will be discussed. The calorimetric approach will be validated using known/classical DNA intercalating and minor groove binding compounds. Although the main focus of the chapter will be analysis of calorimetric data, the data will also be compared to studies on the same systems using ITC-independents approaches such as a gel electrophoresis based topoisomerase I DNA unwinding assays and fluorescence-based ethidium bromide displacement studies.

2. Methods and materials

2.1. Isothermal titration calorimetry

In general, calorimetric titrations were carried out on a MicroCal VP-ITC (MicroCal Inc., Northampton, MA), an instrument specifically suited for studying biomolecular interactions. The MicroCal VP-ITC is a highly sensitive microcalorimeter that operates on a power compensation method, whereby heat exchange processes occurring in a sample cell is compared to a reference cell as the instruments keeps the two cell temperatures identical. This results in exothermic processes yielding negative (less than zero) peaks as the instrument decreases the power (μcal/s) supplied to the sample cell relative to the reference cell, while endothermic processes yield positive (greater than zero) peaks as the instrument increases the power supplied to the sample cell compared to the reference cell. The intensity of each peak corresponds to the quantity of the heat exchange. The data was analyzed using the Origin 7.0 software provided by the manufacturer. Experiments were typically run at either 25-30 °C in MES00 buffer (1 × 10^{-2} M MES (2(N-morpholino) ethanesulfonic acid) containing 1 × 10^{-3} M EDTA, with the pH adjusted to 6.25 with NaOH) for runs involving calf thymus DNA (ctDNA, ultrapure, Invitrogen). Due to the relative instability of the shorter DNA sequence (particularly the AT-rich sequence), experiments using the [poly(dAdT)]$_2$ and [poly(dCdG)]$_2$ sequences (Midland Certified Reagents, Midland , TX) were done in MES40 (i.e., MES00 with 40 mM NaCl). Note, the MES00 buffer was selected for the ctDNA studies due to its low concentration of salt; this would presumably promote stronger binding interactions which would yield more intense peaks and thus better signal/noise ratios. Typically, either 5 or 12 μL of the drug solution (typically 5-7 × 10^{-5} M) was injected into a buffered solution of DNA (typically 10-15 × 10^{-6} M in bp, 1.4 mL) over 20-24 s at 240 s intervals using a 250 μL syringe rotating at 300 rpm. The initial delay (hold period before injections) was set at 240 s. Before use, samples were degassed at 20 °C using the ThermoVac accessory (provided by MicroCal Inc.). During the isothermal titration experiments, all injections manifested in a peak that corresponded to the decrease in the power (μcal/s) supplied to keep the temperatures of the sample and reference cells (containing either water or MES buffer) the same for each injection and represented the heat given off. Note, in all cases, titration peaks corresponded to negative power compensation resulting from exothermically driven processes. In each case, response signals were corrected for the small heat of dilution associated with the titration of the drug into the MES buffer. The heat of dilution for titrating MES buffer into DNA was found to be negligible. The heat released (i.e., area associated with negative peaks) on binding of the drug to DNA sites was directly proportional to the amount of binding. A binding isotherm of heat released (kcal/mol of injectant) versus the molar ratio ([drug]/[DNA] in bp) was constructed and the data fitted by non-linear least square fitting analysis to an appropriate model.

2.2. Topoismerase I DNA unwinding assay

Typically, 0.24 µg of supercoiled pUC19 plasmid DNA was incubated with human topoisomerase I (Topo I) enzyme (Invitrogen) for 5 min at 37 ºC. An appropriate amount of the compound of interest was then added (all except for the first two tubes, which serves as controls) and the reaction mixture incubated for a further 1 h at 37 ºC. After incubation, the reaction was terminated using 0.5% SDS and 0.5 mg/mL proteinase K. Both the enzyme and compound of interest was then extracted using a mixture of phenol:chloroform:isoamyl alcohol (25:24:1). The remaining DNA sample was then run on an agarose gel (1%) at 75 V for 3 h, stained with ethidium bromide for 45 min and photographed.

2.3. Ethidium bromide displacement assay

A solution of ethidium bromide (EtBr, 5×10^{-6} M, 1.0 mL) was pre-incubated with ultrapure calf thymus DNA (1×10^{-5} M in base pairs, 1.4 mL) obtained from Invitrogen. at room temperature (22-23 °C) for 15 min in MES00 buffer, pH 6.3. Aliquots of exactly 3 µL of the compound (7×10^{-5} M) were then titrated into the EtBr-DNA solution and the change in fluorescence measured (Photon Technology International fluorometer), after 3 min incubation periods (excitation 545 nm and emission 595 nm). The addition of 3 µL aliquots was continued until the DNA was saturated (i.e., no further change in fluorescence due to EtBr displacement). [28,36] Control experiments showed that the compounds (free or DNA-bound) had no significant background fluorescence at the excitation (545 nm) and emission (595 nm) wavelengths of EtBr.

3. Results and discussion

3.1. Using relative binding affinity for AT- vs GC-DNA to evaluate binding mode

In order to validate the approach of using relative preferences for AT vs GC to ascertain the preferred DNA binding mode, several known/classical DNA binding compounds were investigated using ITC. These include two compounds known to bind DNA via the minor groove, distamycin A and berenil, (Figure 4) and two compounds known to bind DNA via intercalation (ethidium bromide, normally regarded as the classical DNA intercalator, and daunomycin) (Figure 5).[2,3,6]

Isothermal titration calorimetric data for distamycin, berenil, daunomycin and ethidium bromide binding to the AT- and GC-rich sequences are shown in Figure 6. As can be seen from the raw data, both minor groove binders distamycin A and berenil show a strong preference for the AT-rich sequence relative to the GC-rich sequence. In fact, ITC signals for each compound binding to the GC-rich sequence was found to be negligible, showing only background signal that was associated with the heats of dilution when the compound was titrated into the cell buffer. Binding constants found for distamycin A and berenil binding to the AT-rich sequence were $2.20\pm0.4 \times 10^7$ M^{-1} and $1.76\pm0.3 \times 10^6$ M^{-1}, respectively.

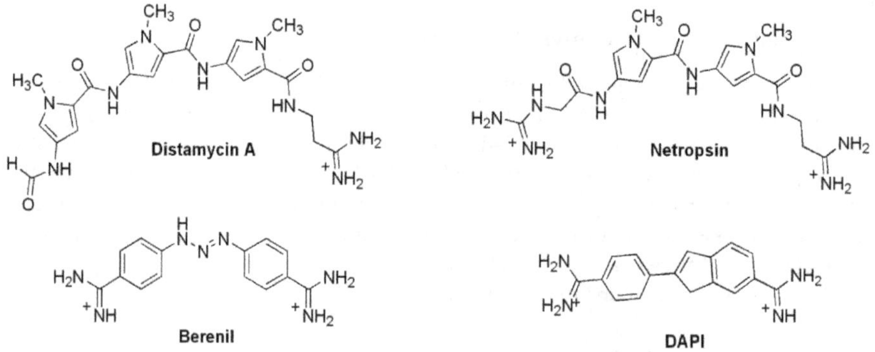

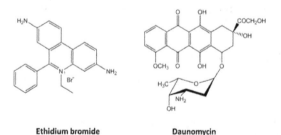

Figure 4. Structures of some common DNA minor groove binding compounds.

A different result was observed with the classical DNA intercalator, ethidium bromide and the known chemotherapeutic DNA intercalator, daunomycin. The isothermal calorimetric data for ethidium bromide and daunomycin showed binding to both the AT- and GC-rich sequences and indicated no significant preference for either sequence. Binding constants obtained for the AT-rich and GC-rich sequence were $1.78\pm0.5 \times 10^5$ M^{-1} and $3.38\pm 0.8 \times 10^5$ M^{-1}, and $2.93\pm0.63 \times 10^6$ M^{-1} and $3.24\pm0.60 \times 10^5$ M^{-1}, for ethidium bromide and daunomycin, respectively.

Ethidium bromide   **Daunomycin**

Figure 5. Structures of two common DNA intercalators.

The results observed for distamycin A, berenil, ethidium bromide and daunomycin are consistent with both distamycin A and berenil binding via the minor groove, since each compound showed a significant preference for the AT-rich sequence, while as expected, ethidium bromide and daunomycin bind DNA via intercalation, since neither exhibited a significant preference. This is suggested from the fact that the minor groove in the GC-rich sequence is partially blocked by the protruded 2-NH₂ group of guanine, preventing a compound that uses the minor groove for DNA binding to be blocked.[6] This is not the case for the AT-rich sequence. On the other hand, a compound such as ethidium bromide and daunomycin which intercalates into DNA by sliding between adjacent base pairs, will essentially be unimpeded from binding to either the AT or GC-rich sequences. The reported binding modes for distamycin A, berenil, ethidium bromide and daunomycin herein are also consistent with the wealth of literature reports on the binding mode for all four compounds, thus validating our approach.[2,3,6,8,10,37,38]

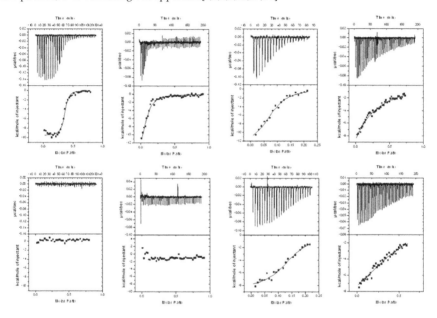

Figure 6. Calorimetric data for the titration of 60 µM of the compounds (from left to right): distamycin A, berenil, daunomycin and ethidium bromide into 15 µM of AT-rich DNA (top), GC-DNA (bottom) at 30 °C. Binding isotherms (heat change vs drug/DNA molar ratio) were obtained from the integration of raw data and fitted to a "one-site" model

4. Binding of the NDI derivatives to DNA using ITC

As was mentioned earlier, the NDI class of compounds is an excellent model system to study DNA binding interactions especially since it offers a useful platform for the syntheses of many homologous series. These molecules are threading intercalators in which side chains on either side of the main intercalating moiety provides the potential for specific recognition sites on the DNA.[12-19] The specific roles of a variety of substituents will be studied with a focus on identifying differential contributions from each moiety. A quaternary amino group will also be incorporated into each NDI side chain to provide electrostatic interaction with the negatively charged DNA backbone. The NDI derivatives in this chapter (Figure 7 and 8) were synthesized by Dixon and coworkers and have three main motifs.[26,36]

Ring Size: Compounds that contains a ring (N-methyl pyrrolidine or N-methyl piperidine) at the distal end of the side chain, as well as possessing different ring size. To date, the effect of ring size on intercalator-DNA interaction has been mostly unexplored. We have studied two homologous types of NDI that differ by a single carbon with five- vs six-membered heterocyclic rings. These are at identical distances from the main intercalating moiety. The rings are non-aromatic and are not expected to stack with the DNA bases. However, they differ in steric bulk which should have implications during binding. One could predict that **NDI-3** will show relatively lower binding affinity than **NDI-4**, however, the increase in bulkiness might have only kinetic consequences.[26] We are interested in determining whether these substituent variations might have an effect on both the preferred DNA binding mode adopted by these compounds, and consequently their relative DNA binding affinity. We also compare the effect of having a cyclic structure in the side chain vs. acyclic alkyl substituents.

Linker length: Insights into the effect of changing the linker length for two sets of NDI derivatives (acyclic aliphatic and cyclic aliphatic substituents) will be discussed. In both sets of compounds, the side chain linker length differ by one carbon (ethyl vs propyl). This means the quaternary amino group (present in all the NDI compounds) is one carbon further from the main intercalating core for the propyl linker. For the acyclic aliphatic derivatives, we compare the trimethyl-propylamino (**NDI-1p**) and dibutylmethyl-propylamino (**NDI-2p**) derivatives (that are one carbon further from the main intercalating core) to the trimethyl-ethylamino (**NDI-1e**) and dibutylmethyl-ethylamino (**NDI-2e**) derivatives. For the cyclic aliphatic compounds, the ethyl-linker-containing compound, **NDI-3**, is compared to the propyl-linker-containing **NDI-5**. Given the difference in steric bulk of the cyclic aliphatic compared to the acyclic derivatives, there may be steric consequences. We will also be able to gain insights into acyclic vs. cyclic substituent effects on DNA binding.

Substituent length/size: In order to gain additional insights into the role of the side chain size, an analysis of the DNA binding characteristics of NDI compounds that differ in the size and side chain linker-length of their alkyl-amine side chain will also be done. As the length and size of the substituent increases, so does the steric bulk. Of course, hydrophobicity also increases with substituent size. We seek to investigate the effects of steric bulk and hydrophobicity on DNA binding of these derivatives. Hydrophobicity has been reported to be a significant driving force in DNA binding interactions with binding increasing with hydrophobicity.[2,3] We have investigated the relative importance of this factor using a model NDI series in which size/steric contributions should also be a factor. Both hydrophobicity and molecular size increases along the series. If hydrophobicity is the predominant driving force, then one might expect binding to increase with size/hydrophobicity. However, if a size/steric effect dominates, binding should decrease.

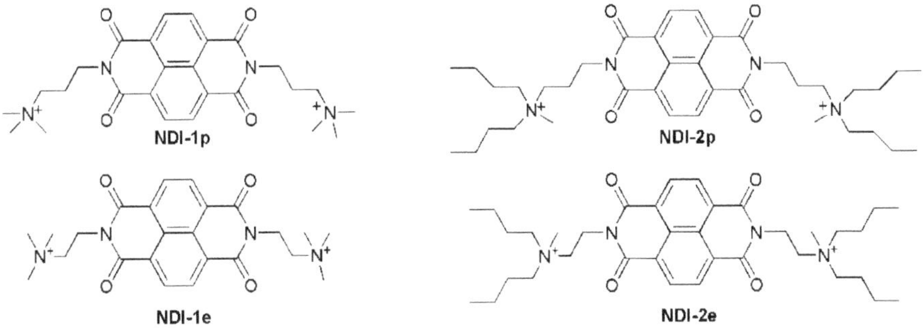

Insights into the Relative DNA Binding Affinity and Preferred Binding Mode of Homologous
Compounds Using Isothermal Titration Calorimetry (ITC)

115

Figure 7. Representative structure for the acyclic NDI derivatives, showing the ethylamino (ethyl
linker) side chain derivatives [**NDI-1e** (bottom, left) and **NDI-2e** (bottom, right)] and the propylamino
(propyl linker) derivatives [**NDI-1p**,(top, left) and **NDI-2p** (top, right)].

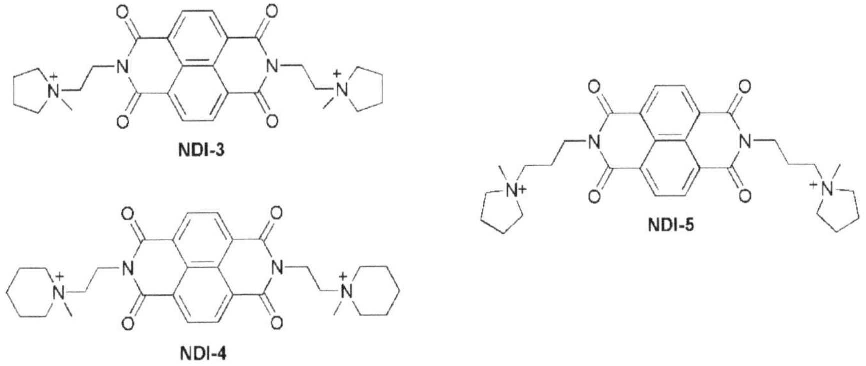

Figure 8. Structures for the cyclic ethylamino NDI derivatives [**NDI-3** (top, left) and **NDI-4** (bottom, left)]
and the cyclic propylamino derivatives **NDI-5** (right). Both **NDI-3** and **NDI-5** contain a side chain N-
methyl pyrrolidine five-membered ring, while **NDI-4** contains a six-membered N-methyl piperidine ring.

5. Effect of the side chain ring size and linker length

In general, the inclusion of a cyclic component in the side chain resulted in a biphasic raw
calorimetric data for each cyclic NDI compound binding to DNA (**Figure 9**). The raw
calorimetric data for the cyclic compounds binding to ctDNA were best defined by a model
that assumes two types of binding sites (K_1, K_2) and argues for the involvement of at least
two different types of binding modes for the compounds with ring-containing substituents.
This biphasic binding mode has been reported by us for larger members of an acyclic
substituent NDI series and will be briefly discussed below.[36] In general, the higher
binding constant (K_1) for the cyclic NDI derivatives was in the order ($\sim 10^7$-10^8 M^{-1}), while a
lower binding constant (K_2) was in the order of ($\sim 10^6$ M^{-1}) for compounds possessing the N-
methyl pyrrolidine ring (**NDI-3** and **NDI-5**) binding to ctDNA. The DNA binding constant
for the N-methyl piperidine derivative (**NDI-4**) showed strong but significantly lower
binding constants compared to the N-methyl pyrrolidine derivatives. Calorimetric data for

the two compounds that differed only in ring size (N-methyl pyrrolidine vs N-methyl piperidine) showed that **NDI-3** (N-methyl pyrrolidine substituent) exhibited larger binding constants (K_1 = 1.17± 0.3 x 10^8 M^{-1}, K_2 = 5.6±0.65 x 10^6 M^{-1}) as compared to larger **NDI-4** (K_1=1.70 ±0.4 x 10^7 M^{-1} and K_2=3.26 ±0.54 x 10^6 M^{-1}) when binding to ctDNA. Thus both binding constants were lower for the larger N-methyl piperidine derivative. Given that **NDI-4** possesses a more bulky N-methyl piperidyl substituent suggest that steric hindrance may play a role here. Studies on a series of NDI containing acyclic substituents also found two binding constants; one binding constant was found to be as a result of intercalation, while the other was found to via a non-intercalative mode, presumably via the DNA minor groove.[36] Assuming that the two binding modes found for the cyclic substituents here are similar (given the similarities between the two sets of compounds), the two binding modes found here for the cyclic derivatives are presumed to also be via intercalation (lower binding constant , K_2) and minor groove binding (higher binding constant, K_1). In which case, **NDI-4** with its larger more bulky substituent may find difficulty in sliding itself through adjacent base pairs. This is of course a requirement for intercalation. Furthermore, given that these compounds possess two substituents on either side of the main intercalating moiety (i.e., threading), one substituent must "thread" through DNA base pairs if it is to adopt an intercalating geometry. Since both binding constant decrease for the N-methyl piperidine derivative, the second binding mode (i.e., presumed to be via the minor groove) is also affected sterically.

According to the calorimetrically determined binding constants, the linker length did not appear to have significant role for these cyclic side chain containing derivatives since **NDI-5** (ethylamino/ethyl linker) and **NDI-3** (propylamino/propyl linker) both had very similar binding constants for both the higher and lower binding sites (K_1 = 1.08 x 10^8 M^{-1}, K_2 =5.1±0.72 x 10^6 M^{-1} and K_1 = 1.17± 0.3 x 10^8 M^{-1}, K_2=5.6±0.65 x 10^6 M^{-1}, respectively). It therefore appears that the size of the cyclic substituent plays a greater role than the substituent linker in determining the DNA binding affinity.

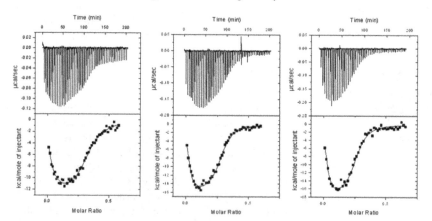

Figure 9. Calorimetric data for the titration of 60 µM **NDI-4** (left), **NDI-3** (middle) and **NDI-5** (right) into 12.5 µM of ctDNA at 30 °C. Binding isotherms (heat change vs drug/DNA molar ratio) were obtained from the integration of raw data and fitted to a "two-site" model.

6. NDI binding mode determination via AT vs GC preference of the cyclic NDI derivatives

Calorimetric studies were carried out to evaluate preferences for AT vs GC-rich sequences, in an effort to detect a possible minor groove binding mode, implied by the above result (**Figure 10**). In general, the cyclic NDI derivatives possessing the ethylamino linker (**NDI-3** and **NDI-4**) exhibited a roughly two-fold preference (2.0x for **NDI-3**, 2.4x for **NDI-4**) for the AT-rich sequence relative to the GC-rich sequence (**Table 1**). The difference in affinity for the AT- vs GC-rich sequence is similar to at least one of the acyclic substituent NDI compounds (a dipropylmethyl ethylamino side chain) reported in an earlier study (see section on the acyclic derivatives below), and which was suggested to have a second minor groove binding mode.[36] We therefore suggest here that the cyclic NDI derivatives **NDI-3** and **NDI-4** does have a minor groove binding mode. It is interesting to note that **NDI-4** showed a slightly greater preference for the AT-rich DNA sequence compared to **NDI-3**, implying a greater involvement of minor groove binding for **NDI-4**.

The cyclic derivative with the propylamino linker (**NDI-5**) exhibited even less of a preference (~1.4x). However, the difference between the **NDI-5** binding constant for AT vs GC-rich sequences could be considered as the same within experimental error. This result may imply that there is a greater contribution from non-intercalative binding from the cyclic ethylamino derivatives relative to the propylamino derivatives. This result is somewhat similar to what was observed in the series of acyclic substituent NDI derivatives. However, given the small differences in AT vs GC-sequences, this would warrant additional studies to confirm.

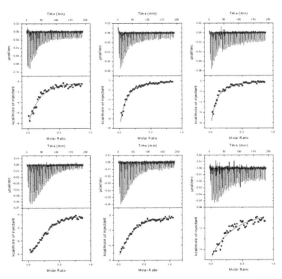

Figure 10. Calorimetric data for the titration of 60 μM **NDI-5** (left), **NDI-3** (middle) and **NDI-4** (right) into 15 μM of AT-rich DNA (top), GC-DNA (bottom) at 30 °C. Binding isotherms (heat change vs drug/DNA molar ratio) were obtained from the integration of raw data and fitted to a "one-site" model.

7. DNA binding mode determination using ITC-independent approaches

Two additional approaches were also utilized to determine the binding mode involved for the compounds in this study. These were a topoisomerase I DNA unwinding assays (topo assay) and ethidium bromide (EtBr) displacement studies. A brief description of the two techniques is in order. Briefly, the topo assay exploits the ability of topoisomerase I enzyme to relax supercoiled DNA, such as the plasmid pUC19 used in all our studies.[39,40] Under the conditions of our topo assay, supercoiled plasmid pUC19 DNA is first relaxed by using excess topoisomerase I enzyme and then is exposed to the compound under study. After extraction of the compound and enzyme, a compound that was bound via intercalation will cause re-supercoiling of the plasmid DNA. Re-supercoiling is due to the change in DNA linking number that accompanies relaxation by the topoisomerase enzyme and occurs to the extent to which the intercalator molecule was initially bound.[39,40] An intercalating molecule will perturb the DNA such that the DNA will unwind, causing the topoisomerase enzyme (which is present in excess) to relax the DNA, thus changing the linking number. The extent to which DNA unwinding occurs will be dependent upon the extent to which DNA binding occurs, thus the minimum concentration needed to cause complete re-supercoiling will be indicative of how much compound was initially bound and thus the relative binding affinity. Conversely, minor groove binders should not induce appreciable re-supercoiling due to negligible DNA unwinding upon binding, and negligible change in DNA linking number.

With the EtBr displacement assays, EtBr, a known intercalator is first bound to DNA, occupying its intercalative sites. The compound of interest is then added to determine whether it is able to displace EtBr from its intercalative sites. Displacement is monitored by a decrease in EtBr–DNA fluorescence.[28,36,41] It is well established that the fluorescence yield of EtBr is enhanced significantly when it binds to DNA. This occurs as EtBr occupies its intercalative sites between bases in the DNA molecule. However, in the presence of another intercalator, there is competition for a limited/defined number of intercalation sites. As the other intercalator molecules are added, they begin to displace EtBr from these intercalative site, increasing the amount of free (unbound) EtBr. This is usually observed as a decrease in EtBr-DNA fluorescence.

Both the topo assay and ETBr displacement assays has been used by our group, as well as other groups, to determine DNA binding mode of DNA binding compounds.[28,35,36,41,42] To validate the topo assay approach, we have run assays on several known DNA binding compounds. These include the classical DNA intercalator, EtBr, and known minor-groove binding compounds such as distamycin A, berenil. **Figure 11** shows representative topo assay for EtBr and berenil.[42] As is expected, the classical DNA intercalator, EtBr, was able to elicit significant re-supercoiling back to the levels of the control (lane 1), whereas, the known minor groove binding compound was unable to do so, even at the high concentrations. In fact, essentially no re-supercoiling was observed for berenil, confirming its known minor groove binding mode. Similarly, we have done validation studies of our EtBr displacement assay, by running studies on DNA binding compounds in which their binding modes have already been established (e.g., distamycin A and berenil). As expected, none of the compounds known to be minor groove binders were able to cause appreciable displacement of EtBr from its intercalative sites, consistent with these compounds binding to non-intercalative sites (**Table 1**). However, the intercalating molecules were able to displace ethidium bromide effectively, as was evident by the significant decrease in EtBr-DNA complex fluorescence.

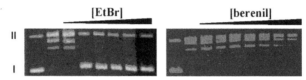

Figure 11. Topo I assay of of the classical DNA intercalator **EtBr** (left) and the known minor groove binder **berenil** (right) using 5 units of the topoisomerase enzyme. From left of each gel, lanes 1 contain only DNA (no compound nor topoisomerase) and serve as controls. Lanes 2 contain DNA and topoisomerase, but no compound. Remaining lanes contain DNA, topoisomerase and increasing concentrations of compound (taken from [42]).

8. Binding mode determination of cyclic NDI derivatives via ITC-independent approaches

When topo assays were done on the NDI derivatives containing the cyclic amino side chains (**NDI-5**, **NDI-3**, and **NDI-4**), each compound was able to cause re-supercoiling, indicating that intercalation is indeed involved in the binding of each compound to DNA. This was not surprising since NDI compounds are known to bind to DNA via intercalation.[17-19] However, **NDI-3** was better able to elicit re-supercoiling than **NDI-5**, which was in turn better than **NDI-4**. That is, while **NDI-3** was able to cause complete re-supercoiling of our plasmid DNA at ~6 μM, **NDI-4** requires >10 μM for complete re-supercoiling (**Table 1**). This suggests that the binding of **NDI-3** involves more of an intercalative mode than either **NDI-5** or **NDI-4** and is consistent with what was observed in the ITC studies for these compounds described above. That is, the strength of the lower binding constants (K_2) was in the order **NDI-3**>**NDI-5**>**NDI-4**. The lower binding constant (K_2 in this report), has been found to be that of the intercalative binding mode for a similar series of NDI.[36] It appears that the bulkier N-methyl piperidine is either sterically hindering intercalation, or forcing **NDI-4** into a more non-intercalative binding mode, while **NDI-5**, with its propylamino linker, exhibits lower affinity for the DNA as compared to **NDI-4**. The lower binding affinity associated with the propylamino linker will be addressed later.

The behavior of the cyclic substituent NDI compounds in the ITC studies and topo assays were also consistent with our EtBr displacement studies which showed that **NDI-3** was better able to displace EtBr from its intercalative sites; thus **NDI-3** caused a greater decrease in EtBr fluorescence compared to **NDI-4** (**Table 1**). Our EtBr displacement assays also showed that **NDI-5** was able to displace EtBr to the same extent as **NDI-3**, suggesting that both have a similar intercalative strengths. Again, this is consistent with what we observed in the ITC and topo assay studies described above. That is, **NDI-5** and **NDI-3** having very similar K_2 (ITC), and both eliciting re-supercoiling of the plasmid DNA at roughly similar concentrations.

Compound	K_b (ctDNA) (10^6 M^{-1}) (ITC)[a]	K_b (AT) (10^6 M^{-1}) (ITC)[b]	K_b (GC) (10^6 M^1) (ITC)[b]	Topo assay (10^{-6} M)[c]	EtBr displacement Assay $(\Delta F/\mu L)$[d]
distamycin A	---	2.20 ± 0.4	---	---	6
Berenil	---	1.76 ±0.3	---	---	31
EtBr	---	0.18±0.05	0.34±0.08	2	---
daunomycin	---	2.9 ±0.6	3.24±0.6	---	---
NDI-1e	15±3	1.11±0.27	1.17±0.10	3.5	400
NDI-2e	78±23 3.9±1.1	1.38±0.15	0.38±0.09	>6.7	358
NDI-1p	1.22±0.16	10.1 ±0.7	---	3	---
NDI-2p	0.57 ± 0.2	8.7 ±0.4	---	5	---
NDI-3	117±30 5.66±0.65	0.5±0.09	0.25±0.05	6	949
NDI-4	17.0±4 3.26±0.54	0.39±0.08	0.16±0.04	>10	777
NDI-5	104± 35 5.10±0.72	1.16±0.24	0.85±0.09	>6	1030

[a] MES00 buffer, pH 6.25
[b] MES40 buffer, pH 6.25.
[c] Minimum concentration required for complete re-supercoiling.
[d] Decrease in EtBr fluorescence per μL of compound added.
Data for acyclic **NDI-#e** series are from reference [36].
Data for the acyclic **NDI-#p** series are from reference [42].

Table 1. Representative DNA binding affinity data for the compounds in this study.

9. Effect of the length/size of the substituent and linker length (Ethyl vs propyl)

As was reported by us, data obtained from calorimetric measurements show that the length/size of the substituent plays a significant role in both the preferred binding mode and relative binding affinity of the compounds of these studies.[36] The compounds of this study showed tight binding to DNA with values of K_b between 10^5 to 10^8 M^{-1}, presumably dependent on their preferred mode of binding to DNA. Figure 12 shows the calorimetric data for the four acyclic NDI derivatives (with ethylamino side chain linkers) binding to ctDNA. In that report, we found only a single type of binding constant (binding mode) for the smallest compound in the series (containing a trimethyl-ethylamino side chain).[36] This

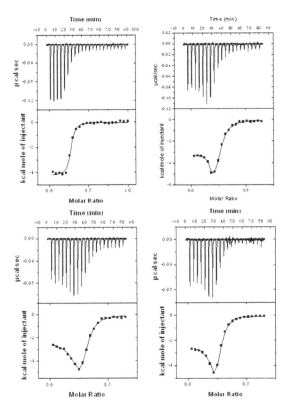

Figure 12. Calorimetric data (raw) for the acyclic ethylamino derivatives binding to ctDNA. In each case, 70 µM of the NDI was titrated into ctDNA (12.5 µM) at 30 °C. Data is shown for the trimethyl ethylamino derivative **NDI-1e** (top, left), diethylmethyl ethylamino derivative (top, right), dipropylmethyl ethylamino derivative (bottom, left), and the dibutylmethyl ethylamino derivative **NDI-2e** (bottom,right). Binding isotherm (heat change vs drug/DNA molar ratio) was obtained from the integration of raw data and fitted to either a "one-site" model (**NDI-1e**) and a "two-site" model (all others). The plot for **NDI-1e** and **NDI-2e** were taken from [36].

was indicated by a single-phased binding isotherm that was well-defined by a one-site binding model. Larger members of the ethylamino series (diethylmethyl-, dipropylmethyl- and dibutylmethyl-ethylamino substituents) adopted two binding modes; a lower affinity binding mode between 3-4 x 10^6 M^{-1} and an additional higher affinity binding mode of between 31 - 78 x 10^6 M^{-1}.[36] This was indicated by a biphasic binding isotherm that was fitted well to a two-site model; one site associated with intercalation and the other associated with minor groove binding. If we compare the results found for the smallest compound in that study, with that of the smallest compound in another study done by us with a similar NDI series with propylamino linker instead,[42] we find that only a single type of binding mode and binding constant (**NDI-1e**, K = 15±3 x 10^6 M^{-1} and **NDI-1p**, K = 1.2

$\pm 0.16 \times 10^6$ M^{-1}) is found for the smallest member of the series whether the side chain is ethylamino or the one-carbon longer, propylamino substituent.[36,42] However, whereas larger members in the acyclic ethylamino series exhibited a dual binding mode, neither compound in the acyclic propylamino series (referred to as **NDI-1p** or **NDI-2p** in this chapter) was found to exhibit more than one binding mode. Additionally, in the ethylamino series, we observe that the relative binding affinity trend for the ethylamino series increased with substituent size. However, this feature was not observed in the propylamino series (one carbon longer on both sides of the main intercalating moiety), since **NDI-2p** with its dibutylmethyl-propylamino substituent exhibited a lower binding constant ($0.57 \pm 0.17 \times 10^6$ M^{-1}) compared to smaller homolog (**NDI-1p**) which had a binding constant of ($1.2 \pm 0.16 \times 10^6$ M^{-1}), a binding mode attributed to intercalative binding.[42] It is also clear that DNA binding affinity was in general greater for the ethylamino derivative, although some of this difference may be attributed to slightly different experimental conditions used in the two studies. It therefore implies that this small structural difference may (1) enable an additional mode of binding, i.e., a linker length that is one carbon shorter resulted in an additional binding mode, as well as (2), enhance the DNA binding mode by greater than an order of magnitude. An explanation for this could be that steric effects may dominate for the propyl-amino series, resulting in lower DNA binding, especially for the larger members, while hydrophobic and binding mode preferences may be dominant in the ethyl-amino series. The propyl amino derivatives are of course longer especially since the additional carbon linker is on both sides of the molecule, given that these are threading compounds. The longer (more dangling) molecular structure may make it more difficult to thread through adjacent base pairs. However, in the case of the ethylamino series, the solution for the larger substituents appear to be adoption of an additional DNA binding mode. Hydrophobic contributions may also play a role.

10. Comparison of binding mode for NDI derivative with ethyl vs propyl linker using topo assay

Comparing the two series with different linker-length (i.e., ethylamino vs propylamino derivatives), it is also interesting to note that generally higher concentrations of the ethylamino derivatives were required for re-supercoiling, despite having higher binding constants as determined by ITC.[36,42] A striking example of this is seen from the fact that more than 6.5 uM of **NDI-2e** ($K_1 = 78 \pm 23 \times 10^6$ M^{-1} and $K_2 = 3.9 \pm 1.1 \times 10^6$ M^{-1}) was required for supercoiling, while the corresponding propylamino derivative **NDI-2p** with a significantly lower binding affinity ($K = 0.57 \pm 0.17 \times 10^6$ M^{-1}) required only 5 uM. Again, some of this may also be attributed to different experimental conditions. For example, a greater excess of the topoisomerase enzyme was used in the assays for the ethylamino series. However, this factor alone cannot account for the lack of associated re-supercoiling ability given the disproportionately higher DNA binding constants for the ethylamino derivatives. Overall, a side by side comparison of the topo assay results for the two series (ethylamino vs propylamino) suggests that the ethylamino derivatives displays relative re-supercoiling capabilities that are less than expected based on their significantly higher binding affinities. Since the ability to elicit re-supercoiling is primarily based on an intercalative ability, this argues for a greater involvement of non-intercalative binding for ethylamino derivatives relative to their propylamino counterparts.

11. Binding of the chalcogenoxanthylium derivatives to DNA

In an effort to further corroborate our DNA binding characterization approach used for the NDI derivatives discussed above using a different/independent homologous series, we will also briefly describe DNA binding studies of a homologous series of chalcogenoxanthylium derivatives to DNA, reported by our group.[35] The chalgenoxanthylium derivatives in this study were synthesized by Detty and coworkers and have been implicated as potential candidates for therapy against blood-borne pathogens.[27,31-33,35].

Using this independent system as a comparison, we have also found that the results obtained from ITC were consistent with that found using topo assay and EtBr displacement studies. These studies have found that the nature of the substituent attached to the main xanthylium core plays a directing role in the preferred binding mode and accompanying DNA binding affinity.[35] While some of the compounds bind to DNA either through intercalation or via the minor groove, some exhibited mixed-binding modes.[35] Excerpts from the DNA binding studies for selected chalcogenoxanthylium derivatives (**Figure 13**) will now be discussed.

In that report, ITC studies suggested that both the 9-substituent and the identity of the chalcogen play a role in the preferred binding mode and ultimately, the relative DNA binding constant.[35] With a 9-2 thienyl substituent attached to the main xanthylium core (e.g., 2-Se), there appeared to be a preference for intercalation. This was implied from the fact that compounds containing the 9-2 thienyl substituent showed no preference for the AT-rich sequence, a feature that would be typical for a minor-groove binder. The 9-2 thienyl also bound to calf thymus DNA with lower affinity as compared to the 9-phenyl derivatives (e.g.,1-Se).[35] DNA intercalators are known to have lower DNA binding affinity as compared to minor-groove binders,[2] so this result may be due to a greater contribution from minor groove binding (i.e., less contribution from intercalation) with the 9-phenyl series. In addition to exhibiting a 2-3 higher binding constant compared to the corresponding 9-2 thienyl derivative, the 9-phenyl series exhibited a slight preference (2-3 times) for binding to [poly(dAdT)]$_2$ as compared to the [poly(dGdC)]$_2$. Here again, a possible minor groove binding was implied, since it is known that compounds that bind solely to the DNA minor groove generally show a preference for binding to AT-rich sequences relative to GC-rich sequences due to the occlusion from the GC-rich minor groove by the protruded 2-NH$_2$ group of guanine.[6] As mentioned for the NDI series discussed earlier, it is expected that compounds that bind both via the DNA minor groove and by intercalation (i.e., mixed binding modes) will show a factor of <10 preference for AT-rich sequences, depending on the relative contribution from intercalation (i.e., the difference will be less as contributions from intercalation increases). The chalcogenoxanthylium derivative bearing a 9-(2-thienyl-5-diethylcarboxamide) substituent (compound 10) exhibited the strongest preference for the [poly(dAdT)]$_2$ sequence. In fact, compound 10 showed essentially no binding to the [poly(dGdC)]$_2$ sequence, while binding to [poly(dAdT)]$_2$ with a K of 2.3 ±0.4 x 10^6 M^{-1}.[35]

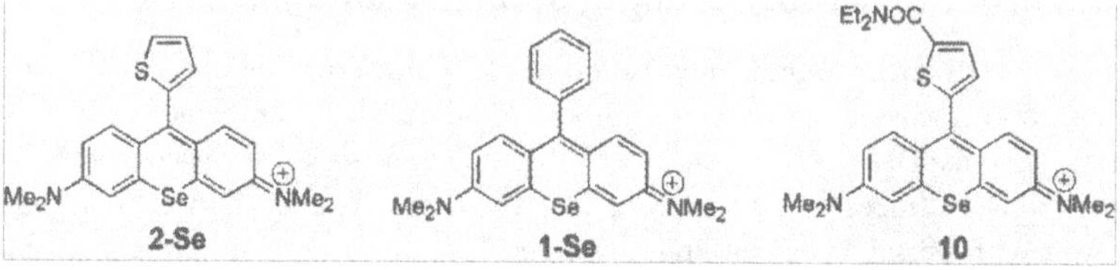

Figure 13. Structures of selected chalcogenoxanthylium derivatives reported in [35]. The 9-2 thienyl derivative (**2-Se,** left) shown bind mostly via intercalation, while **1-Se** derivative (middle) is a mix-binder, and compound **10** binds primarily via the DNA minor groove.

12. Binding mode determination of chalcogenoxanylium derivatives via ITC-independent approaches

As was done for the NDI series discussed earlier, several independent (non-ITC) studies (ethidium bromide displacement and topo assay) were also carried out on the chalcogenoxanthylium derivatives in this study.[35] This was done in an effort to gain additional insights into the preferred DNA binding mode suggested by ITC.

Results from topo assays have been reported by us.[35] These results were in general consistent with the ITC studies on these compounds. We will now report new EtBr data on chalcogenoxanthylium derivatives discussed in this chapter that supports both ITC and topo assay studies.

Further evidence for the preferred DNA binding modes were also observed during ethidium bromide displacement assays on selected members of the chalcogenoxanthylium compounds binding to DNA. These were the seleno derivatives from the 9-2 thienyl series (**2-Se**), the 9-phenyl series (**1-Se**), and compound 10 (suggested to have primarily a non-intercalative binding from the ITC studies). While compound **2-Se** and **1-Se** were both able to cause dislodgement of ethidium bromide from DNA, **2-Se** was markedly better able to do so (decrease in fluorescence per µL of compound added was: **2-Se** = 711, **1-Se** = 581, compound 10 = 350. Considering that part of the change is fluorescence for the compounds was due to accompanying dilution during the titration, we see here that the order of intercalative ability is **2-Se**>**1-Se**>**10**. This order mirrors the results from both ITC and topo assay which showed that **2-Se** was a better intercalator than **1-Se**, which was in turn better than compound 10. This implies that **2-Se** is a stronger intercalator than **1-Se**, consistent with both the ITC and topo assay studies. Compound 10 caused relatively small decreases in ethidium bromide fluorescent (less than any of the NDI derivatives in this study) indicating that it is not a potent displacer of ethidium bromide from its intercalative sites, suggesting that compound **10** is not a strong intercalator, again consistent with the minor groove binding mode implied by both the ITC and topo assay. Given the higher binding constant found for **1-Se** relative to **2-Se** using ITC, if **1-Se** was primarily a DNA intercalator, it would exhibit a greater ability (compared to **2-Se**) to dislodge the classical DNA intercalator EtBr from its binding sites. The fact that it did not, strongly supports the idea that the binding of **1-Se** to DNA involves other binding modes. Also, the fact compound **10** showed little ability to dislodge ethidium bromide from DNA, while having the highest binding constant (as determined by ITC studies in an earlier study [35]), supports the idea of compound **10** involving significant non-intercalative DNA binding (presumably, via the minor-groove).

13. Conclusions

In this chapter, we have shown how ITC can be successfully used to characterize both the preferred DNA binding mode for series of compounds, as well as their relative DNA binding affinity. For this, we have selected two homologous series of compounds; series of symmetrical NDI threading intercalators in which the side chains are mandatorily involved in DNA binding, and a series of chalcogenoxanthilum derivatives. Both classes of compounds have been shown to have biological activity.

While the homologous NDI derivatives in this study all exhibit DNA intercalative abilities, the substituent on either side of the main intercalating core does play a significant role in determining whether or not additional modes are adopted. This occurs because these compounds require a threading geometry when intercalating between DNA base pairs, i.e., there is a necessity for the side chain to "thread" DNA. The side chains are therefore forced to direct DNA binding. We have found that the cyclic (non-aromatic) substituent at the distal end of a side chain play a significant role in both the DNA binding affinity and the preferred mode of binding. Larger ring sizes face steric barriers and have lower DNA binding affinity. The larger rings may however force additional (non-intercalative) binding modes to be involved. Additional studies may be needed to fully understand the full effects of ring size. Future studies may involve attachment of aromatic rings instead of non-aromatic rings in this study. Having flat aromatic rings on the substituent may enhance site recognition and DNA binding due to the ability to stack. We have also found that even a small modification in the linker length in NDI side chain play a significant role during binding of NDI derivatives of acyclic aliphatic side substituents to DNA. In fact, on comparing side chains with an ethyl linker vs those with a propyl linker, it was found that the ethyl linker could enhance DNA binding by more than an order of magnitude. Possession of the ethyl linker also enabled an additional DNA binding mode of higher affinity. The NDI scaffold therefore represent a versatile template for the design of many promising derivatives with enhanced DNA affinity and have implications in the rationale design of DNA binding compounds with improved site recognition capabilities.

Using an independent system for comparison, the approach of using ITC to study binding to both ctDNA and AT vs GC-rich sequences, was shown to be an efficient and consistent approach in the determination of relative DNA binding affinity and preferred DNA binding mode. The ITC studies were well corroborated by ITC-independent studies such as topo assays and EtBr displacement studies, thus exhibiting the efficacy of our approach.

Author details

Ruel E. McKnight

Department of Chemistry, State University of New York at Geneseo, 1 College Circle, Geneseo, NY, USA

Acknowledgement

The author is very grateful to Professors Dabney Dixon and Michael Detty for providing the naphthalene diimide and chalcogenoxanthylium compounds, respectively, for this study. I would also like to acknowledge the very diligent students who have contributed to this work over the years (Douglas Jackson, Luke Marr, Kevin Siegenthaler, Eric Reisenauer, Sadia Sahabi, Shivani Polasani, Bilgehan Onogol, Manuel Pintado, Aaron Gleason, and James Keyes).

14. References

[1] Bailly C, Colson P, Hénichart J-P, Houssier C (1993) The different binding modes of Hoechst 33258 to DNA studied by electric linear dichroism. Nucleic Acids Res. 21:3705-3709.

[2] Chaires, JB (1997) Energetics of Drug-DNA Interactions. Biopolymers 44: 201-215.

[3] Haq I (2002) Thermodynamics of Drug-DNA Interactions. Arch. Biochem. Biophys. 403:1-15.

[4] Barcelo, F.; Capo, D.; Portugal, J. (2002) Thermodynamic characterization of the multivalent binding of chartreusin to DNA. Nucleic Acids Res. 30:4567-4573.

[5] Tse WC, Boger DL (2004) A Fluorescent Intercalator Displacement Assay for Establishing DNA Binding Selectivity and Affinity. Acc. Chem. Res. 37:61-69.

[6] Ren J, Chaires JB (1999) Sequence and Structural Selectivity of Nucleic Acid Binding Ligands. Biochemistry 38:16067-16075.

[7] Denny WA (2002) Acridine Derivatives as Chemotherapeutic Agents. Curr. Med. Chem. 9: 1655-1665.

[8] Chaires, JB (1996) Molecular Recognition of DNA by Daunomycin Advances in DNA Sequence Specific Agents. 2:141-167.

[9] Chen AY, Liu LF (1994) DNA Topoisomerases: Essential Enzymes and Lethal Targets. Annu. Rev. Pharmacol. Toxicol. 34, 191-218.

[10] Pilch DS, Kirolos MA, Liu X, Plum GE, Breslauer KJ (1995) Berenil [1,3-bis(4'-amidinophenyl)triazene] Binding to DNA Duplexes and to a RNA Duplex: Evidence for Both Intercalative and Minor Groove Binding Properties. Biochemistry 34:9962-9976.

[11] Haq I, Jenkins T, Chowdhry B, Ren J, Chaires JB (2000) Parsing Free Energies of Drug-DNA Intercalation. Methods Enzymol. 323: 373-405.

[12] Hampel SM, Sidibe A, Gunaratnam M, Riou JF, Neidle S (2010) Tetrasubstituted Naphthalene Diimide Ligands with Selectivity for Telomeric G-Quadruplexes and Cancer Cells. Bioorg. Med. Chem. Lett. 20:6459-6463.

[13] Sato S, Hirano A, Takenaka S (2010) Selective Immobilization of Double Stranded DNA on a Gold Surface Through Threading Intercalation of a Naphthalene Diimide having Dithiolane Moieties. Analytica Chimica Acta. 665:91-97.

[14] Lee J, Guelev V, Sorey S, Hoffman DW, Iverson BL (2004) NMR Structural Analysis of a Modular Threading Tetraintercalator Bound to DNA. J Am Chem Soc. 126:14036-14042.

[15] Gianolio DA, Segismundo JM, McLaughlin LW (2000) Tethered Napthalene-based Intercalators for Triplex Stabilization. Nucleic Acids Res. 28: 2128-2134.

[16] Liu ZR, Hecker KH, Rill RL (1996) Selective DNA Binding of (N-alkylamine)-Substituted Naphthalene Imides and Diimides to G+C-rich DNA. J Biomol Struct Dyn.14:331-339.

[17] Tanious FA, Yen SF, Wilson WD (1991) Kinetic and Equilibrium Analysis of a Threading Intercalation Mode: DNA Sequence and Ion Effects. Biochemistry 30:1813-1819.

[18] Wilson, W. D. DNA Intercalators. In DNA and Aspects of Molecular Biology; Kool, E. T., Ed.; Elsevier: New York, 1999; pp 427-476.

[19] Yen S, Gabbay E, Wilson WD (1982) Interaction of Aromatic Imides with Deoxyribonucleic Acid. Spectrophotometric and Viscometric Studies Biochemistry, 21:2070-2076.

[20] Sato S, Kondo H, Takenaka, S, (2006) Linker Chain Effect of Ferrocenylnaphthalene Diimide Derivatives on a Tetraplex DNA Binding. Nucleic Acid Symposium Series 50:107-108.

[21] Rusling D, Peng G, Srinivasan N, Fox K, Brown T, (2009) DNA Triplex Formation with 5-Dimethylaminopropargyl Deoxyuridine. Nucleic Acid Res 87:1288-1296.

[22] Cuenca F, Greciano O, Gunaratnam M, Haider S, Munnur D, Nanjunda R, Wilson W, Neidle S (2008) Tri- and tetra-substituted Naphthalene Diimides as Potent G-Quadruplex Ligands. Bioorg. Med. Chem. Lett.18:1668–1673.

[23] Laronze-Cochard M, Kim Y-M, Brassart B, Riou J-F, Laronze J-Y, Sapi J (2009) Synthesis and Biological Evaluation of Novel 4,5-Bis(dialkylaminoalkyl)-Substituted Acridines as Potent Telomeric G-Quadruplex Ligands, Eur. J. of Med. Chem. 44:3880–3888.

[24] Gonzalez, V, Hurley, L (2010) The c-Myc NHE III: Function and Regulation, Annu. Rev. Pharmacol. Toxicol. 50:111-129.

[25] Luedtke, N, (2009) Targeting G-Quadruplexes with Small Molecules. Chimia 63: 134-139.

[26] Steullet V, Dixon, DW (1999) Self-Stacking of Naphthalene bis(dicarboximide) Probed by NMR. Perkin Trans. 2:1547-1558.

[27] Wagner, S, Skripchenko, A, Donnelly, D, Ramaswamy, K, Detty, M. (2005), Bioorg. Med. Chem. 13:5927-5935.

[28] McKnight, RE, Ye M, Ohulchanskyy, TY, Sahabi S, Wetzel, BR, Wagner, SJ, Skripchenko A, Detty MR (2007) Synthesis of Analogues of a Flexible Thiopyrylium Photosensitizer

for Purging Blood-Borne Pathogens and Binding Mode and Affinity Studies of their Complexes with DNA. Bioorg. Med. Chem. 15:4406-4418.

[29] Wainwright M (1998) Photodynamic Antimicrobial Chemotherapy. J. Antimicrob. Chemother 42:13-28.

[30] Dougherty T, Gomer C, Henderson B, Jori G, Kessel D, Korbelik M, Moan J, Pend Q (1998) Photodyanamic Therapy. J. Natl. Caner. Inst. 90:880-905.

[31] Detty MR, Gibson SL, Wagner SJ (2004) Current Clinical and Preclinical Photosensitizers for use in Photodynamic Therapy. J. Med. Chem 47, 3897-3915.

[32] Wagner, SJ, Skripchenko A, Cincotta L, Thompson-Montgomery D, Awatefe H (2005) Use of a Flexible Thiopyrylium Photosensitizer and Competitive Inhibitor for Pathogen Reduction of Viruses and Bacteria with Retention of Red Cell Storage Properties. Transfusion 2005, 45, 752-760.

[33] Calitree B, Donnelly D, Holt J, Gannon M, Nygren C, Sukumaran D, Autschbach J, Detty M. (2007) Tellurium Analogues of Rosamine and Rhodamine Dyes: Synthesis, Structure, [125]Te NMR, and Heteratom Contribution to Excitation Energies. Organometallics 26:6248-6257.

[34] Gibson S, Hilf R, Donnelly D, Detty M (2004) Analogues of Tetramethylrosamine as Transport Molecules for and inhibitors of P-Glycoprotein-Mediated Multi-Drug Resistance. Bioorg. Med. Chem. 12:4625-4631.

[35] McKnight, RE, Onogul B, Polasani SR, Gannon MK, Detty MR (2008) Substituent Control of DNA Binding Modes in a Series of Chalcogenoxanthylium Photosensitizers as Determined by Isothermal Titration Calorimetry and Topoisomerase I DNA Unwinding Assay. Bioorg. Med. Chem. 16:10221-10227.

[36] McKnight, RE, Reisenauer, E, Pintado, MV, Polasani, SR and Dixon, DW (2011) Substituent Effect on the Preferred DNA Binding Mode and Affinity of a Homologous Series of Naphthalene Diimides, Bioorg. Med. Chem. Lett. 21:4288-4291.

[37] Barcelo F, Portugal J, (1993) Berenil Recognizes and Changes the Characteristics of Adenine and Thymine Polynucleotide Structures. Biophys. Chem. 47:251-260.

[38] Remata D, Mudd C, Berger R, Breslauer K (1993) Thermodynamic Characterization of Daunomycin-DNA Interactions: Comparison of Complete Binding Profiles for a Series of DNA Host Duplexes. Biochemistry 32:5064-5073.

[39] Pommier, Y, Covey J-M, Kerrigan D, Markovits J, Pham R (2007) DNA Unwinding and Inhibition of Mouse Leukemia L1210 DNA Topoisomerase I DNA Intercalators. Nucleic Acids Res. 15:6713-6731.

[40] Dziegielewski J, Slusarski B, Konitz A, Skladanowski A, Konopa (2002) Intercalation of Imidazoacridinones to DNA and its Relevance to Cytotoxic and Antitumor Activity. J. Biochem. Pharmacol. 63:1653-1662.

[41] Boger DL, Fink BE, Brunette, SR, Tse WC, Hedrick, MP (2001) A Simple, High-Resolution Method for Establishing DNA Binding Affinity and Sequence Selectivity. J. Am. Chem. Soc. 123:5878-5891.

[42] McKnight, R. E.; Gleason, A. B.; Keyes, J. A.; Sahabi, S. (2007) Binding Mode and Affinity Studies of DNA Binding Agents Using Topoisomerase I DNA Unwinding Assay. Bioorg. Med. Chem. Lett. 17:1013-1017.

Chapter 6

Applications of Calorimetric Techniques in the Formation of Protein-Polyelectrolytes Complexes

Diana Romanini, Mauricio Javier Braia and María Cecilia Porfiri

Additional information is available at the end of the chapter

1. Introduction

1.1. Formation of the protein-polyelectrolyte complex

Polyelectrolytes are flexible-chain polymers containing subunits with negative or positive charges. These compounds form soluble or insoluble complexes with proteins with opposite electrical charge.

The different equilibriums present in a solution of protein and polyelectrolyte are shown in figure 1.

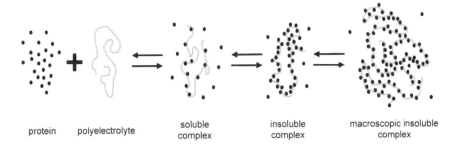

protein   polyelectrolyte   soluble complex   insoluble complex   macroscopic insoluble complex

Figure 1. Formation of the protein- polyelectrolyte complex (binary complex).

As can be seen, when a protein interacts with a polyelectrolyte, a soluble complex is formed containing few molecules of the protein. As more molecules of protein interact with the polyelectrolyte, the complex becomes insoluble. Finally, the particles of insoluble complex start to interact with each other to form the macroscopic insoluble complex.

Besides this model of formation of the insoluble complex, there is another model in which the formation of the insoluble complex requires the presence of an inorganic polyion (figure 2).

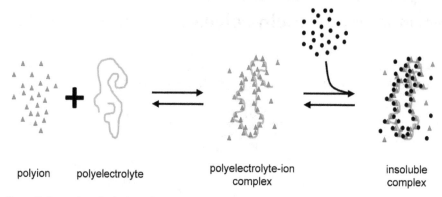

| polyion | polyelectrolyte | polyelectrolyte-ion complex | insoluble complex |

Figure 2. Formation of polyelectrolyte-ion-protein complex (ternary complex).

First, the polyelectrolyte interacts with the polyion and forms a soluble complex with charges oppose to those in the protein. Then, the protein interacts with the complex to form the insoluble complex.

In both cases, the formation and solubility of the complex depends on the pH and the ionic strength of the medium [1], the density of charges in the protein and the polyelectrolyte, the molecular weight and the concentration of the polyelectrolyte [2,3]. Various studies have been directed to understand the formation of these complexes in aqueous medium [4-6]. Equation 1 shows how the density of charges (σ) on the surface of the protein and the polyelectrolyte is affected by the pH and the ionic strength of the medium.

$$\sigma = \frac{\partial \sigma}{\partial pH}(pH - pI) \tag{1}$$

Mattison *et. al.* postulated an equation that correlates the density of charges in the protein (σ) and the polyelectrolyte (ξ) with the Debye-Hückel constant (κ) that is highly dependent on ionic strength (a is a constant that depends on the protein-polyelectrolyte system) [7].

$$\xi\sigma \cong a \cdot \kappa \tag{2}$$

The conditions of the medium determine whether the soluble or the insoluble complex is formed, or if the complex is dissociated.

1.2. Stoichiometry of the protein-polyelectrolyte formation

When studying the interaction of a protein and a polyelectrolyte, it is interesting to know the minimum quantity of protein requires forming the maximum quantity of complex per polyelectrolyte unit. This value is called stoichiometry of the complex (*e*) and it is usually represented as the moles of polyelectrolyte per mol of protein, mass of polyelectrolyte per mass unit of protein, etc.

The stoichiometry of a protein-polyelectrolyte complex might be over or below 1 as shown in figure 3.

A complex with a stoichiometry below 1 contains more protein molecules than polyelectrolyte molecules. Usually, this kind of complex is insoluble, while a complex with a stoichiometry over 1 might be soluble or insoluble. As can be seen in figure 3, a complex with a stoichiometry below 1 can be turned into one with stoichiometry over 1 by adding polyelectrolyte to the medium. Of course, the effect can be reverted by adding protein.

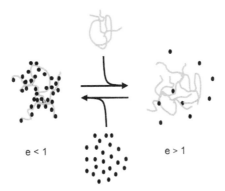

Figure 3. Stoichiometry (e) of a protein-polyelectrolyte complex.

1.3. Biotechnological applications of the protein-polyelectrolyte complex

In biotechnology, it is interesting to use insoluble protein-polyelectrolyte complexes with e < 1 since they can be used to purificate and concentrate industrial-interest enzymes [8], to immobilized enzymes in bioreactors or scarefolds.

1.3.1. Bioseparation of proteins from a complex mixture

The development of Biotechnology has allowed the used of enzymes in the production of food, drugs and in many others industries. At the same time, Genetic Engineering and Molecular Biology has allowed the expression of proteins in bacteria and superior microorganisms; however, some proteins must still being isolated from its natural sources due to complex post-traductional modifications that occur during the synthesis of the proteins. In both cases, the protein of interest is in a media containing many other biomolecules and inorganic compounds that need to be separated from the protein before applied it to any industrial process.

Most purification protocols consist on many steps: the first ones have the aim to concentrate the protein of interest and to obtain a high recovery; while the last steps of the protocol are expected to yield a high purification factor.

Precipitation of enzyme-polyelectrolyte insoluble complexes is a very useful technique to apply at the beginning of purification protocols since it requires simple equipment and so is very easy to scale up; needs low concentrations of the polyelectrolyte since it interacts with high affinity with the proteins; and can be based on a wide variety of polyelectrolytes, both natural and synthetic, positively or negatively charged. An important aspect of this technique is that the enzyme usually retains its tertiary structure and its catalytic activity. Even more, usually it is more stable in the presence of the polyelectrolyte [9,10].

1.3.2. Enzyme immobilization

The immobilization of enzymes is a process by which the protein is attached to a solid matrix, synthesized using a polyelectrolyte, in order to enhance the stability of the structure of the protein and to reuse enzyme [11]. Enzymes immobilized, in comparison with enzymes in solution, are more robust and resistant to environmental changes.

Immobilization can be performed by physical or chemical methods. The first results in weak interactions between the enzyme and the solid support and includes adsorption on a water-insoluble matrix and gel entrapment or micelles [12,13]. Chemical methods generate covalent bonds or electrostatic interactions between the enzyme and a water-insoluble support forming reticulated or single-chain particles. Insoluble complexes allow to immobilized enzymes by entrapment in polyelectrolyte solid particles, micelles or by covalent linkage to the support using carboimide as coupling.

The protein-matrix systems are widely used in bio-reactors for industrial process mainly because of the possibility of recycle the enzyme. Bio-reactors are very useful for the synthesis of organic compounds; since immobilized enzymes reduced the steps of the process, enhance the purity of the final products and allow stereo-selective synthesis. These systems also can be applied on the production of micro/nanocapsules for the delivery of proteins (or drugs). In this case, the use of polyelectrolytes sensitive to environmental conditions allows the releasing of the enzyme (or drug) molecules in different parts of the body [14].

1.4. Characterization of the protein-polyelectrolyte complex

The formation of the protein-polyelectrolyte complex can be studied by spectroscopic and calorimetric techniques.

Spectroscopy assays based on turbidimetric measurements allow knowing the effect of pH, ionic strength, time and temperature on the amount of insoluble complex formed. Phase diagrams, turbidimetric titrations and kinetic assays must be performed in order to evaluate the best conditions to obtain the maximum quantity of insoluble protein-polyelectrolyte complex [15;16].

Study the stability of the enzyme when it is part of the complex is also important to understand how the polyelectrolyte affects its catalytic activity since it is expected to use the enzyme in an industrial process. Fluorescence, UV-visible and circular dichroism spectroscopy are very useful techniques to analyze the structure of the protein but also must be performed experiments to study the enzymatic activity.

Although these techniques are helpful, they do not give an idea of the nature of the interaction between the protein and the polyelectrolyte. Calorimetric techniques such as differential scanning calorimetry and isothermal titration calorimetry allow studying the thermal stability of the enzyme in the presence of the polyelectrolyte and the nature of the interaction between the enzyme and polyelectrolyte, respectively.

2. Research course and methodology

In order to study the formation of insoluble complexes between proteins and polyelectrolytes it is necessary to carry out different methodologies in a sequential way.

2.1. Titration curves at different pH in binary systems

The formation of the insoluble polyelectrolyte-protein complexes can be followed by means of turbidimetric titration of the protein with the polyelectrolyte, or vice versa. Taking into account the isoelectric point of the protein and the pK value of the polyelectrolyte, the pH of the medium should be selected so that they have opposite net charge.

Figure 4. shows the turbidimetric titration curve obtained for two systems with different behavior: hyperbolic and sigmoid. Turbidity is usually measured as the absorbance (Abs) at 420 nm.

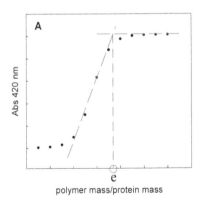
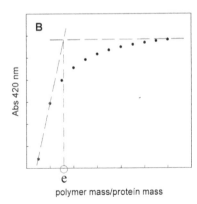

Figure 4. Determination of the polyelectrolyter/protein mass ratio when saturation is reached in a **A**- sigmoid and **B**- hyperbolic graph

These graphs demonstrate a saturation behavior with different mechanism of complex formation, and allow us to determine the *stoichiometric polyelectrolyte/protein ratio "e"*, which is defined as the minimal ratio in which the protein precipitates as an insoluble complex. The value "e" is calculated from the plot at the lowest concentration of polyelectrolyte necessary to get the saturation. This value is important because it allows us to determine the minimal amount of polymer needed to fully precipitate the protein, and can be expressed as the number of moles of protein bounded per mol of polyelectrolyte.

2.2. Titration curves at different pH in ternary systems

In ternary systems, the polyelectrolyte forms an insoluble complex with an anion, which associates proteins with opposite net charge. The mixture polyelectrolyte-anion behaves as a pseudo polyampholyte with a characteristic isoelectric point.

The formation of insoluble complexes between the polymer and the anion can be examined by turbidimetric titration of the anion with the polyelectrolyte. When these curves reach the saturation it suggests a complete precipitation of the complex.

The precipitation of polyampholyte -protein complexes is driven by coulombic forces, which are highly dependent on protein and polyampholitic isoelectric pH values [5;17;18]. So, precipitation begins at a critical pH where the attractive forces have just overcome electrostatic repulsion.

2.3. Phase diagrams in binary systems

Phase diagrams, also called solubility curves, show the range of pH in which the complexes are soluble or insoluble. It means that they provide information about the pH of higher interaction between the components and the optimum pH for precipitation and dissolution of the complexes.

To obtain these diagrams, a polyelectrolyte/protein mixture at a ratio close to "e" is titrated with acid or alkali and the turbidity of the medium is measured after pH variation.

Figure 5. shows an schematic phase diagram in a binary system.

2.4. Phase diagrams in ternary systems

Classical polyampholytes have both anionic and cationic groups in their molecules. However, the aqueous solution of any polyelectrolyte may behave as a polyampholyte provided there is a small ion with multiple electrical charges (two or more) in the medium which interacts with the opposite charge of the polyelectrolyte to form a pseudo polyampholyte. Under these conditions, it is possible to find a pH interval where the pseudo complex behaves as an ampholyte.

To obtain the phase diagrams, a mixture with fixed polyelectrolyte/anion ratio is titrated with alkali or acid, and the turbidity of the medium is measured as the absorbance at 420 nm after pH variation.

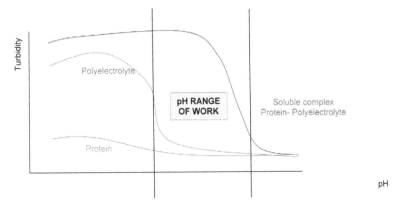

Figure 5. Phase diagram, turbidity vs. pH, for protein (—), polyelectrolyte (—) and binary system (—).

Figure 6 shows a schematic phase diagram between a cationic polyelectrolyte and an anion, in different polymer/anion ratios. Figure 6 A shows a pH range where the turbidity of the medium drastically increases. Each curve has a trapezoidal shape with a plateau, and the height of the trapezium depends on the polyelectrolyte concentration. The pHs that correspond to the edges of the trapezium are the critical pH values, at which the transition from complete dissolution to precipitation occurs. The lower critical pH is usually called acidic, and the higher one is called basic. It is remarkable that the transitions from complete solubility to precipitation occur at the same critical pHs independently of the polyelectrolyte concentration.

Also, phase diagrams can be represented as in figure 6.B. This diagram represents the behavior of the polyelectrolyte-anion complex by filled and open circles: filled circles are drawn at the pH of non-zero absorbance whereas the open circles at the zero absorbance values of the solution.

As mentioned above, insoluble polyelectrolyte-anion complexes behave as an ampholyte. This can be used to precipitate cationic or anionic proteins depending on the pH of work, as indicated in figure 6.B.

2.5. Effect of ionic strength

The coulombic component in the interaction between proteins and polyelectrolytes is closely related with the presence of charges. So, the ionic strength of the medium can alter the forces involved in the interaction and eventually leads to dissociation of the complexes.

In ternary systems, high ionic strength can also inhibit polymer-anion interaction.

Whatever the system, this inhibition of the formation of the precipitates may be directly proportional to salt concentration. So, the effect of the presence of salt in the systems can be evaluated by turbidimetric titration in the presence of different concentrations of NaCl in the medium.

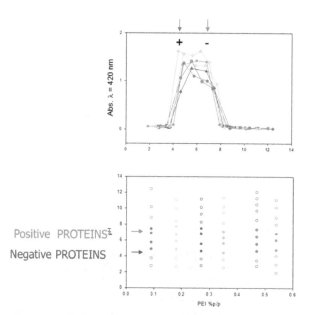

Figure 6. Phase diagram in polyelectrolyte-anion systems. Each color represents a different polyelectrolyte/anion ratio.

2.6. Complex formation kinetics

The interaction between polyelectrolytes and proteins requires time to achieve the maximum quantity of complex (maximum turbidity). Thus, it is necessary a kinetic study by which the turbidity of a mixture polyelectrolyte-protein is measured over time. So, a solution of the polyelectrolyte is added to a solution of the protein, at the pH of precipitation and in an appropriate ratio, and absorbance at 420 nm is measured over time. Finally, a plot of turbidity vs. time is made and the time required to reach the maximum quantity of turbidity is obtained [19].

2.7. Conformational and enzymatic evaluation of the protein in the complex

Several investigations have reported that polymers stabilize the catalytic activity in a variety of enzymes. Besides, it has been suggested that electrostatic interactions between the enzyme and polyelectrolytes play a primary role, also in conformational stabilization [20;21].

2.7.1. Effect of the polyelectrolytes on the far-UV circular dichroism (CD) of proteins

Circular dichroism spectroscopy is a very frequently used technique to evaluate protein conformation in solution. This method is sufficiently simple and allows a rapid determination of protein structure or conformational changes.

In the far-UV region (between 180 and 250 nm) the circular dichroism spectrum provides information one the secondary structure of proteins, due to asymmetrical packing of intrinsically achiral (planar) peptide groups [22].

The effect of polymers on the structure of proteins can be analyzed in terms of its secondary elements. So, far-UV circular dichroism spectra of proteins are recorded in different polymer/protein ratios, with a fixed concentration of protein and varying the amount of polymer. The pH of the medium must be the pH of higher interaction between the protein and the polyelectrolyte.

2.7.2. Effect of the polyelectrolytes on the enzymatic activity of proteins

In order to evaluate the effect of the polyelectrolyte on the enzymatic activity of the protein, enzyme assays are performed at constant protein concentration in the presence of different amounts of polymer. Polyelectrolyte/protein ratios are usually close to the stoichiometry of the complex ("e") or in excess of polymer respect this value.

The stability of the enzyme in the presence of the polyelectrolyte can also be monitored by incubating the mixture polyelectrolyte/enzyme and recording the enzymatic activity over time.

2.8. Precipitation and redissolution of the complexes

After analyzing the conditions of complex formation or dissolution and evaluating the effects of the different variables, we are able to design a methodology of precipitation of the protein with the polyelectrolyte by following the steps shown in figure 7.:

According to this precipitation scheme, an aliquot of the polyelectrolyte is mixed with a solution of the protein, both prepared at the pH of precipitation, to reach a proper polymer/protein ratio. This mixture is incubated the time necessary to form the maximum quantity of insoluble complex and centrifuged to obtain a solid precipitate. Then, the supernatant is separated and the precipitate redissolved in buffer at the pH of dissolution of the complex. Finally, in order to evaluate the effectiveness of the total process, enzymatic activity is measured in both fractions.

This scheme is successfully applied in many systems [15-20] and allows to obtain a protein with a high recovery and catalytically conserved.

2.9. Calorimetric measurements for polymer-protein complex

2.9.1. Differential scanning calorimetry

Thermal desnaturation of proteins was monitored with a high sensitivity differential scanning calorimeter model VP-DSC from MicroCal Inc. Thermograms were obtained between 20-85°C, at scan rate 25-30ºC/h and at constant pressure of 28 psi. All result were averages of, at least, three independent measurements. Buffer versus buffer baseline scans were determined and subtracted from transition scans prior to normalization and analysis of protein denaturation. Finally, the values of the excess heat capacity were obtained after subtraction of the baseline. The calorimetric data were analysed by using the software ORIGIN 7.0, MicroCal Inc., following the methodology recommended by IUPAC. The parameters obtained from this analysis were: temperature at which maximum heat exchange occurs (Tm), the area under the peak, which represents the enthalpy of transition for reversible process (ΔH_{cal}) and the van't Hoff enthalpy (ΔH_{VH}).

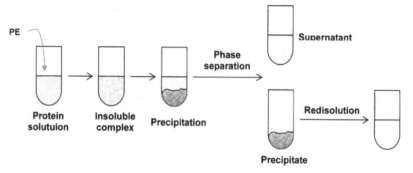

Figure 7. Scheme of precipitation of polyelectrolyte/protein complexes

The evaluation of ΔH_{VH} gives an idea of the mechanism of the unfolding process [23]:

- $\Delta H_{VH} = \Delta H_{cal}$: a two-state process is carried out under equilibrium condition.
- $\Delta H_{VH} > \Delta H_{cal}$: intermolecular cooperation is taking place which may require some degree of molecular association.
- $\Delta H_{VH} < \Delta H_{cal}$: one or more intermediate states of significance in the overall process.

However, in some cases this calorimetric criterion may lead to erroneous conclusion.

2.9.2. Isothermal titration calorimetry

Measurements of the examples presented were performed at 20-25 ºC by using a VP-ITC titration calorimeter (MicroCal Inc. USA). The sample cell was loaded with 1.436 mL of protein solution and the reference cell contained Milli-Q grade water. Titration was carried out using a 300 µL syringe filled with polyelectrolyte solutions. The experiments were performed by adding aliquots of 3-5µL of polyelectrolyte solutions 0.175 % (w/w) to the cell containing the protein solution.

The mathematical model equation selected to fit the ITC data was derived from a model that assumes the polyelectrolyte molecule binding to several protein molecules, all with the same intensity; in other words, the polyelectrolyte was considered as a macromolecule having **n** independent and equivalent sites, all of which have the same affinity constant, K, for the ligand (protein) [24].

The heat associated with the interaction polyelectrolyte-protein (ΔH_b) was calculated by subtraction using the equation 3:

$$\Delta H_b = \Delta H_t - \Delta H_d - \Delta H_{dissol} \tag{3}$$

Where ΔH_t is the total heat associated to each polymer addition, ΔH_d is the heat of dilution of the polyelectrolyte in the buffer in the absence of the protein and ΔH_{dissol} is the heat of polymer dissolution. The heat associated to the dilution of the protein in buffer was negligible. Then ΔH_b was plotted vs polyelectrolyte/protein molar ratio and, by non-linear fitting of these calorimetric curve, the affinity constant (K) for the polyelectrolyte binding to the protein and the number of polymer molecules (n) bound per protein molecule was calculated using the software provided by the instrument.

The resulting data set was fitted using MicroCal ORIGIN 7.0 software supplied with the instrument and the intrinsic molar enthalpy change for the binding, ΔH_b, the binding stoichiometry, n, and the intrinsic binding constant, K, were thus obtained. The equation for determining the heat associated to each injection is:

$$Q = \frac{n \, M_t \, \Delta H_b \, V_0}{2} \left(1 + \frac{1}{nkM_t} + \frac{X_t}{nM_t} - \sqrt{\left(1 + \frac{1}{nkM_t} + \frac{X_t}{nM_t}\right)^2 - \frac{4X_t}{nM_t}} \right) \tag{4}$$

where V_0 is the active volume cell, X_t is the bulk concentration of ligand and M_t is the bulk concentration of the macromolecule in V_0 [25].

The intrinsic molar free energy change, ΔG^0, and the intrinsic molar entropy change, ΔS^0, for the binding reaction were calculated by the fundamental thermodynamic equations 5 and 6:

$$\Delta G^\circ = -R \, T \ln K \tag{5}$$

$$\Delta S^\circ = \frac{\Delta H^\circ - \Delta G^\circ}{T} \tag{6}$$

3. Results

3.1. Turbidimetric titration curves

Figure 8 shows typical hyperbolic titration curves of a protein with a polyelectrolyte. In this case, trypsin (TRP) with poly vinyl sulfonate (PVS) [26]. As can be seen, two important characteristics were observed: at low polymer/protein ratios, absorbance increases with an increase in the polyelectrolyte total concentration and, at high polyelectrolyte/protein ratio, there is a plateau which depends on the medium pH.

The protein/polyelectrolyte molar ratio which corresponds to the situation where the protein has been precipitated as an insoluble complex was calculated from the intersection of a straight line which corresponds to the prolongation of the linear zone of the curve (at low polymer concentration) with a line which gives a plateau.

Trypsin is a serin-protease found in the digestive system. It is used for numerous biotechnological processes. Its isoelectric point is between 11.0 and 11.4 [27]. The pHs selected in the curves were chosen in the range where TRP and PVS have opposite charges.

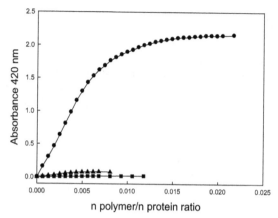

Figure 8. Turbidimetric titration curves of TRP (70μM) solution with PVS (0.25 % w/w) in a medium with phosphate buffer 50mM, pH 3.0 (●), 5.5 (▲) and 7.0 (■). T=20 ºC, [21].

Table 1 shows the molar protein-polyelectrolyte ratio which corresponds to the stoichiometry of the complex formation calculated from the titration curves for the different experiments. These values are important because they allow us to calculate the minimal polymer amount necessary to precipitate the protein in a complete form. The data have been expressed as the number of TRP moles bound per polyelectrolyte mol. Despite the fact that these values were similar, turbidity is much higher at pH 3.00 which suggest a major size of the precipitate particle.

pH	Protein/polyelectrolyte molar ratio
3.00	136 ± 3
5.50	228 ± 21
7.00	147 ± 12

Table 1. TRP/PVS molar ratios at different pHs.

Figure 9 shows titration curves of lysozyme (LYZ) with the polyelectrolyte PVS. LYZ is a basic protein with 19 amino residues, an isoelectrical pH between 11.0 and 11.4 and a molecular mass of 14.3 kDa [28], therefore at the pHs where the turbidimetry titration were assayed the protein has a net positive electrical charge. Formation of LYZ-PVS complex was observed to be influenced by the medium pH, however, at pH 3.1, a minor absorbance maximum value was observed than at pH 5.5, which can be assigned to the loss of the native structure of this protein by influence of the acid medium [26].

Table 2 shows the molar protein-polymer ratios which correspond to the stoichiometry of the complex formation calculate from the titration curves for the different experiments. These values are important because allow to estimate the minimal polyelectrolyte amount needed to precipitate the protein, the data have been expressed as the number of LYZ molecules bound per polyelectrolyte molecule.

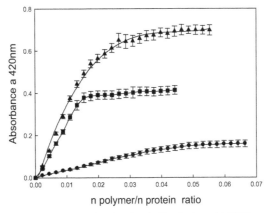

Figure 9. Turbidimetric titration curves of LYZ (0.3mg/mL) solution with PVS in a medium with 50 mM phosphate buffer. pH 5.5 (▲), 7.0 (●), and acetic acid/acetate buffer pH 3.1 (■). T= 20 ℃.

LYZ- PVS	Protein/polyelectrolyte molar ratio
pH 3.1	66
pH 5.5	47
pH 7.0	23

Table 2. Lys/PVS molar ratios at different pHs.

3.2. Turbidimetric titration of ternary complex:

Figure 10 shows turbidimetric titration curves when phosphate (500mM) or citrate (50mM) was titrated by adding a concentrated solution of the polyelectrolyte poly ethylene imine (PEI). Both curves reached a plateau at high polyelectrolyte anion ratios, which suggests a complete precipitation of the complex. It could be seen that the plateau was obtained at a polymer/anion ratio 10 times higher for citrate than for phosphate, suggesting that citrate has a greater precipitation capacity than phosphate. These ternary systems have the capability to precipitate in the absence of protein. Only is required the presence in the medium of the cationic polyelectrolyte (PEI) and a polyanion like phosphate (Pi) or citrate (Cit).

3.3. Phase diagrams of binary systems

Figures 11 show the absorbance dependence (at 420 nm) vs. the pH change by the system LYS with poly acrylate (PAA). These complexes were soluble at basic pH values. At pH lower than 6.50 a significant increase in the turbidity was observed that corresponding to the insoluble complex formation. Similar behavior was reported for the serum albumin titration with anionic polyelectrolyte [7].

The relevance of these phase diagrams are in the possibility to know the insolubility and solubility complex conditions.

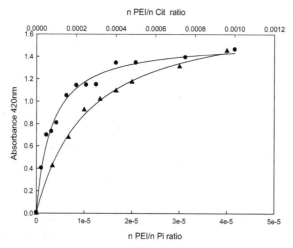

Figure 10. Turbidimetric titration of phosphate (▲) and citrate (●) with PEI. pH 5.5. T= 20° C [6].

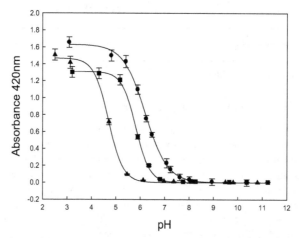

Figure 11. Dependence of the absorbance at 420 nm vs the medium pH at a constant protein-polyelectrolyte molar ratio of LYZ-PAA: (●) 0.0027, (■) 0.0065, (▲) 0.0010. T= 20°C [25].

3.4. Phase diagrams of ternary systems

Figure 12 shows the pH variation effect on the insoluble complex formation obtained for ternary systems PEI/Pi at different PEI/Pi molar ratios [6]. As can be seen, in all curves there is an increase in the turbidity of the medium, reaching a maximum value and then decreasing in the pH interval 3.5-7. Each curve has a trapezoidal shape and the pH values corresponding to the edges of the trapezium are the critical pHs at which the transition from complete dissolution to precipitation occurs.

In this figure it can be identify both critical pHs: 3.5 and 7. At pH=3.5 the net charge of the complex is positive whereas at pH= 7 is negative. On the other hand, transitions from complete solubility of the insoluble complex are independent of the polyelectrolyte concentration.

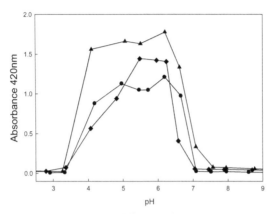

Figure 12. Phase diagram of PEI/Pi systems at different molar ratios.

3.5. Effect of ionic strength

In general, protein/polyelectrolyte insoluble complexes are greatly affected by ionic strength because the molecular mechanism of the interaction is mainly electrostatic in nature. Turbidimetric titrations at pH 7.00 were performed in medium of different ionic strength such as shown Figure. 13. In this case, only 0.1 M of NaCl is enough to avoid formation the insoluble protein-polyelectrolyte complex [26]. This finding may be interesting in the design of an isolation method of protein, allowing in a first step the precipitation by the polyelectrolyte and then the precipitate may be dissolved by NaCl solution addition at low concentration.

Ternary systems like PEI-citrate was dramatically affected by 0.5 for higher ionic strength; in this case, no formation of the insoluble complex was found while the PEI-phosphate system showed to be slightly affected by the NaCl increased concentration. The inhibition of the precipitate formation in both systems was directly proportional to the salt concentration, in agreement with the presence of an important coulombic component in the insoluble complex formation [19].

3.6. Kinetics of the complex formation

In general, the kinetics of complex formation is fast (around 2-10 minutes) [15;17;19]. Figure 14 shows the kinetic studies of different molar ratios of the systems TRP/Eudragit®L100 (EL100). It required less than 2 minutes of incubation to achieve the maximum quantity of complexes. In addition, by increasing the polyelectrolyte concentration increases the turbidity of the system, however the time required achieving the maximum turbidity is independent of the concentration of the molar ratio.

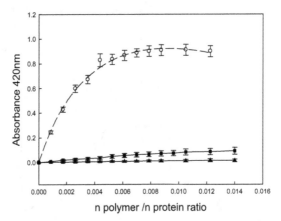

Figure 13. NaCl concentration effect on the turbidity of LYZ-PAA, pH 7.0, NaCl concentration: (O) 0M, (■) 0.1 M and (▲) 0.5M. T= 20ºC.

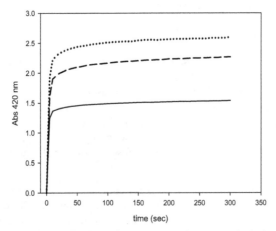

Figure 14. Formation of complex TRP-EL100 through time at three protein/polyelectrolyte molar ratio of TRP/EL100: (——) 32.41, (— —) 16.18, (····) 8.08. [15].

4. Calorimetric techniques of protein-polyelectrolyte complex

4.1. Differential scanning calorimetry by polymer-protein complex

DSC is a useful tool for studying the protein unfolding in which values of excess specific heat capacity (Cp) are obtained as a function of temperature. Two enzymes having different behavior towards charged flexible chain polyelectrolytes are analyzed below.

Lysozyme is a basic protein with 19 amino residues, an isoelectrical pH between 11.0 to 11.4 and a molecular mass of 14.3 kDa. Because LYZ is one of the four proteins whose thermal denaturation is thermodynamically reversible, the equations for systems in thermodynamic equilibrium can be applied to obtain the thermodynamic functions (entropy and enthalpy of unfolding) directly from the thermograms, as described by Privalov [29].

Thermograms of LYS enzyme with PAA and PVS are presented as examples in figure 15 and Table 3 shows the thermodynamics functions and Tm obtained in each case. In these systems DSC measurements demonstrated that the Tm of LYS was not modified by the polyelectrolytes presence only a decrease in the denaturalization heat (ΔH_{cal}) was observed.

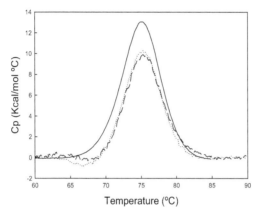

Figure 15. DSC Thermograms of the LYZ in the absence (—) and presence of: PVS (----) and PAA (....) pH 7.00.

	LYZ	LYZ-PVS	LYZ -PAA
ΔH_{cal} (kcal/ mol)	89.4 ± 0.3	72.0 ± 0.3	66.7 ± 0.4
ΔH_{VH} (kcal/ mol)	139.0 ± 0.6	141 ± 0.8	151 ± 1.0
ΔH_{VH} / ΔH_{cal}	1.55 ± 0.05	1.96 ± 0.01	2.26 ± 0.03
Tm (°C)	75.01 ± 0.01	75.33 ± 0.02	75.2 ± 0.1
ΔS (e.u.)	399 ± 3	405 ± 4	405 ± 5

Table 3. Thermodynamic functions obtained for the thermal LYZ unfolding determined by DSC in the absence and presence of the studied polyelectrolytes.

A Tm constant value is a proof that the protein retains its thermodynamic stability in the presence of both polyelectrolytes. However the polymer presence induced a decrease in the area under the curves, in agreement with a diminution of the heat associated to the denaturation process. The unfolding entropic change showed to be not affected by the polyelectrolyte presence, in accordance with the protein retaining its tertiary structure and no important conformational protein change is occurring.

LYZ is a protein which has only one domain with low molecular mass, its thermal unfolding have been describe as reversible, however the capacity of LYZ to associate in aqueous solution it is well known. $\Delta H_{VH}/\Delta H_{cal}$ ratio greater than 1 is an indication of the intermolecular cooperation presence during the thermal unfolding. The increase of this ratio in the polyelectrolytes presence, suggests more cooperative intermolecular process.

Furthermore, the unfolding entropy was not affected in the protein-polymer complexes (LYZ-PVS and LYS-PAA). It indicates that LYZ in complex follows in the same conformational state that LYZ alone.

Trypsin is a serin-protease found in the digestive system. It is used for numerous biotechnological processes. It is a globular protein which has two domains with similar structures [27]. DSC experimental results for enzyme trypsin are demonstrated a two-state transition model at pH 3.00 [30]. Figure 16.A shows DSC thermograms of TRP. Although the ratio $\Delta H_{VH}/\Delta H_{cal}$ is close to 1, however the thermogram clearly shows 3 transitions.

TRP-EL100 complex has a very interesting behavior. As can be seen in Figure 16.B protein thermogram was significantly modified by the polyelectrolyte presence.

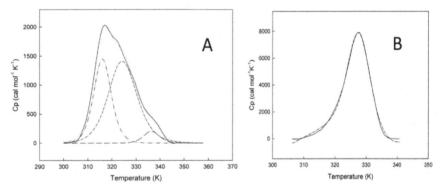

Figure 16. (A) DSC Thermogram of the TRP: (—) experimental data; (---) fit data; (··—) first transition; (- - -) second transition; (— —) third transition. (B) DSC Thermogram of the TRP in the presence of EL100: (—) experimental data; (---) fit data.

	Tm (K)	ΔH_{cal} (kcal/mol)	ΔH_{VH} (kcal/mol)	$\Delta H_{VH}/\Delta H_{cal}$
TRP	320.6 ± 0.1	38.7 ± 0.6	42.2 ± 0.8	1.09
TRP (transitions)	316.1 ± 0.1	13.9 ± 0.1	8.3 ± 3	
	324.4 ± 0.3	22.2 ± 0.2	53.1 ± 3	-
	336.4 ± 0.3	1.8 ± 0.5	-	
TRP-EL100	327.8 ± 0.1	82.0 ± 0.1	81.6 ± 0.1	0.99

Table 4. Thermodynamics functions obtained for the thermal TRP unfolding determined by DSC in the absence and presence of the EL100.

The comparison of the two figures and table evidences two main differences

- Tm of the TRP curve is 320.6 K (47.5°C), whereas for the TRP-EL100 complex the same parameter is 327.8 K (54.6°C). The shift of the Tm of TRP in the presence of the polymer means that the EL100 stabilizes the structure of the enzyme against thermal denaturation.
- The different unfolding model in the absence (3 transitions of independent domains) and in the presence of polymer (two-state model)

4.2. Isothermal titration calorimetry

ITC technique gives the direct heat associated to the complex formation (ΔH), a number of protein molecules bounded to polyelectrolyte molecule (**n**), the affinity constant (**K**). Before performed ITC experiment is important to known which is the number "*e*" obtained by turbidimetric titration because is a good estimation of **n**.

Figure 17 shows the ITC measurements of the LYS titration with PVS and Table 5 summarizes the parameters obtained by two anionic polyelectrolytes (PVS and PAA).

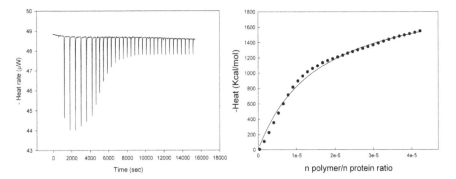

Figure 17. ITC measurements of LYZ with PVS [26].

System	LYZ-PVS	LYZ -PAA
n (protein /polyelectrolyte)	21.2 ± 0.2	294 ± 8
K (M⁻¹)	$2.7\ 10^3$	$5.1\ 10^4$
$\Delta H°$ (kcal/mol)	- 15.2	- 10.0
$\Delta S°$ (e.u.)	-1103	-1033

*The enthalpic change is expressed per mol of protein bound.

Table 5. Thermodynamic and binding parameters of the interaction LYZ-polyelectrolyte from ITC experiments.

The interaction LYZ-PVS and LYZ-PAA is exothermic. The mechanism of bond is carried out between the electrically charged groups of both. The differences found between complexes were the affinity (K) and the number of molecules of protein bonded to polymer molecule. Because the size of the polyelectrolytes are 10-fold larger than the protein, the number of protein molecules bound per polymer molecule is high.

The heat associated to the complex formation were extremely high, but when they are normalized per protein molecule bound to the polyelectrolyte the heat associate yielded 10-15 kcal/mol which is a normal heat amount for a coulombic interaction between two charge groups in solution. These low interaction heats are in agreement with the low NaCl concentration needed to induces the dissolution of the insoluble complex (around 0.1 M) Other important parameters to know are the thermodynamically stability of the protein in the polyelectrolyte presence. It is desirable that the protein retains its tertiary structure.

TRP- EL100 complex is an interesting example. Although the polymer and protein present opposite electrical charge, however the interaction is endothermic.

Figure 18 shows the binding isotherm obtained when consecutive aliquots of EL100 were added to a solution of trypsin [15]. The parameters calculated are summarized in Table 6.

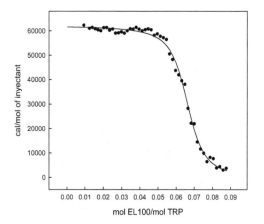

Figure 18. ITC measurements of TRP with EL100.

Binding parameter	Value
n (molar ratio) [mol TRP/mol EL-100]	15.22 ± 0.05
K (affinity constant) $[M^{-1}]$	$9.8\ 10^6 \pm 7.\ 10^5$
ΔH° (kcal/mol)	$62.1\ \pm 0.3$
ΔS°(e.u.)	240 ± 10
ΔG° (kcal/mol)	$- 9.59 \pm 0.05$

Table 6. Binding parameters of the TRP- EL100 interaction from ITC experiments.T= 25°C.

A value of 15 mol of protein per mol of polyelectrolyte was found for the complex EL100-TRP formation. The high value of the affinity constant demonstrated that both molecules interact strongly with each other. The ΔH was normalized per mol of protein; therefore, heat value of 62.14 Kcal/mol of protein is yielded. The positive value of ΔH indicates that the interaction between EL100 and TRP requires consuming of heat form de medium. The $\Delta S°$ value obtained was positive as a result of the increase of the disorder of the system due to release of structured water molecules.

EL100 is a charged polymer which also contains a hydrophobic framework in its linear chain. For such a complicated system it is not clear to what extent non-electrostatic forces contribute to the observed complexation behavior. Besides, the value of $\Delta S°$ was positive indicating that the disorder of the system increased.

ITC experiment performed in presence of NaCl confirmed the results obtained by turbidimetry (data not shown). The values of heat measured during the experiment of titration are similar that obtained when studying the dilution of the polyelectrolyte. This result is indicating that the TRP and the EL100 are not interacting when NaCl 1.00 M is added to the buffer.

Thermodynamic parameters were according to hydrophobic interactions between TRP and EL100. However, ITC and turbidimetric titrations experiments were altered in salt presence. It would demonstrate that the mechanism of interaction between these two molecules involves both hydrophobic and electrostatic interactions.

5. Conclusions

Experimental conditions of charged polyelectrolyte-protein complex formation may be determined by turbidimetric measurements, but are necessary to complement it for calorimetric techniques.

DSC measurements show that the Tm and denaturalization heat of some proteins may increase or not change in the polymer presence and the complex unfolded according to a two-state model.

In general, $\Delta H°$ and $\Delta S°$ of complex formation obtained by ITC have negative when protein and polyelectrolyte are oppositely charged (electrostatic interaction). Nevertheless, the thermodynamic functions can be positive as a result of the interaction between hydrophobic backbone of polymers and aromatic amino acids. Moreover, if ionic strength modifies this insoluble complex formation, a mechanism of interaction may involve both hydrophobic and electrostatic interactions.

The calorimetric techniques (ITC and DSC), turbidimetry and enzymatic activity studies provide useful quantitative information about the interaction of proteins and charged polyelectrolytes in aqueous solution. The knowledge of the nature of this interaction is essential for the application of the complex formation in protocols as proteins isolation strategy, immobilization or in purification of a target protein.

Author details

Diana Romanini, Mauricio Javier Braia and María Cecilia Porfiri
Laboratory of Physical Chemistry Applied to Bioseparation. College of Biochemical and Pharmaceutical Sciences, National University of Rosario (UNR), Rosario, Argentina

Acknowledgement

We thank Prof Watson Loh, Institute of Chemistry, State University of Campinas (UNICAMP), Campinas, SP, BRAZIL for performing DSC and ITC measurements. We also thank Prof. G. Picó, Prof. B. Nerli and Prof. B. Farruggia for useful discussions.

6. References

[1] Kumara, A Srivastavaa A, Yu Galaevb I, Mattiasson B, (2007) Smart polymers: Physical forms and bioengineering applications. Progress in Polymer Science 32: 1205-1237.

[2] Wang, Y.; Gao, Y.; Dubin, P., (1999) Protein Separation via Polyelectrolyte Coacervation: Selectivity and Efficiency, Biotechnology Progress, 12: 356-362.

[3] Arvind, L.; Aruna, N.; Roshnnie, J.; Devika, T. (2000) Reversible precipitation of proteins on carboxymethyl cellulose, Poocess Biochemistry 35: 777-785.

[4] Weinbreck, F.; De Vries, R.; Schrooyen, P.; De Kruif, C.G. (2003) Complex Coacervation of Whey Proteins and Gum Arabic, Biomacromolecules, 4: 293-303.

[5] Gupta V., Nath S., Chand S., (2002) Estimation of proteins in the presence of polyethylenimine. Biotechnol. Lett., 22: 927.

[6] Manzur A, Spelzini D, Farruggia B, Romanini D, Picó G (2007) Polyethyleneimine phosphate and citrate systems act like pseudo polyampholytes as a starting method to isolate pepsin. Journal of Chromatography B. 860: 63-68.

[7] Mattison, K.W.; Dubin, P.L.; Brittain, I.J., (1998) Complex Formation between Bovine Serum Albumin and Strong Polielectrolytes: Effect of Polymer Charge Density, Journal of Physical Chemistry B, 102: 3830-3836.

[8] Cooper C, Dubin P , Kayitmazer A, Turksen S,Current, (2005) Polyelectrolyte–protein complexes Opinion in Colloid & Interface Science 10: 52–78.

[9] Pessoa Jr., A.; Vahan Kilikian, B.;Purificação de Produtos Biotecnológicos, Ed. Manole Ltda., cap. 2. (2005).

[10] Hilbrig F, Freitag R, (2003) Protein purification by affinity precipitation, J Chrom B 790: 79–90.

[11] Arroyo M., (1998) Inmobilized enzymes: Theory, methods of study and applications. Ars Pharmaceutica, 39: 23-39.

[12] Krajewska B., (2004) Application of chitin- and chitosan-based materials for enzyme immobilizations: a review Enzyme Microb Tech 35: 126–139.

[13] Saskia Lindhoud. (2009) Polyelectrolyte Complex Micelles as Wrapping for enzymes – Tesis, 206 p.

[14] Esposito E., Cervellati F., Menegatti E., Nastruzzi C., Cortesi R., (2002) Spray dried Eudragit microparticles as encapsulation devices for vitamin C, Int J Pharm 242: 329–334.

[15] Braia, M Tubio, G Nerli, B Loh W, Romanini, D., (2012) Analysis of the interactions between eudragit® l100 and porcine pancreatic trypsin by calorimetric techniques. Int J Biol Macromol 50: 180-186.

[16] Porfiri M. C., Picó G., Farruggia B., Romanini D., (2010) Insoluble complex formation between alpha-amylase from Aspergillus oryzae and polyacrylic acid of different molecular weight. Proc. Biochem. 45: 1753-1756.

[17] Tsuboi A., Izumi T., Hirata M., J. Xia, P. Dublin E. Kokufuta, (1999) Complexation of Proteins with a Strong Polyanion in an Aqueous Salt-free System Langmuir 12: 6295-6303.

[18] Fornasiero F., Ulrich J., Prausnitz J.,(1999) Molecular thermodynamics of precipitation. Chem. Eng. Process 38: 463-475.

[19] Patrickios, C, Sharma, L, Armes, S, Billingham, N. (1999) Precipitation of a Water-Soluble ABC Triblock Methacrylic Polyampholyte: Effects of Time, pH, Polymer Concentration, Salt Type and Concentration, and Presence of a Protein. Langmuir 15: 1613-1620.

[20] Foreman T, Khalil M, Meier P, Brainard J, Vanderberg L, Sauer N (2001). Effects of charged water-soluble polymers on the stability and activity of yeast alcohol dehydrogenase and subtilisin carlsberg. Biotechnol Bioeng 76: 241–246.

[21] Braia, M. , Porfiri, M.C., Farruggia, B., Picó G., Romanini, D. (2008) Complex formation between protein and poly vinyl sulfonate as a strategy of proteins isolation. Journal of Chromatography B, 873: 139-143.

[22] Fasman G. D. (1996) Circular dichroism and the conformational analysis of biomolecules. Plenum press 738p.

[23] Sturtevant J, (2001) Biochemical applications of differential scanning calorimetry. Annu. Rev. Phys. Chem. 38 463-488.

[24] Jha N, Kishore N., (2009) Binding of streptomycin with bovine serum albumin: Energetics and conformational aspects.Thermochim. Acta 482: 21-29.

[25] Kim W., Yamasaki Y., Kataoka K., (2006) Development of a Fitting Model Suitable for the Isothermal Titration Calorimetric Curve of DNA with Cationic Ligands. J. Phys. Chem. B 110: 10919-10925.

[26] Romanini D, Braia M, Giatte Angarten R, Loh W, Picó G, (2007) Interaction of lysozyme with negatively charged flexible chain polymers. J Chrom B, 857: 25-31.

[27] Beynon R, Bond J.S., Proteolytic Enzymes, Practical Approach, Oxford University Press, 2001.

[28] Aschaffenburg R., Blake C., Dickie H., Gayen, S., Keegan R., Sen A. (1980). The crystal structure of tortoise egg-white lysozyme at 6 Å resolution .Biochim Biophys Acta 625: 64-71.

[29] Privalov P.L..(1979) Stability of Proteins Small Globular Proteins. Adv. Protein Chem. 33: 167-241.

[30] Santos A. Santana M., Gomidea, F. Miranda A, Oliveira, J., Vilas Boas F., Vasconcelos, A., Bemquerer M. ,Santoro M., (2008) Physical-chemical characterization and stability study of α-trypsin at pH 3.0 by by differential scanning calorimetry Int. J. Biol.Macromol. 42: 278–284.

Chapter 7

Thermodynamic Signatures of Macromolecular Complexes – Insights on the Stability and Interactions of Nucleoplasmin, a Nuclear Chaperone

Stefka G. Taneva, Sonia Bañuelos and María A. Urbaneja

Additional information is available at the end of the chapter

1. Introduction

Nucleoplasmin (NP) is a nuclear chaperone that mediates chromatin remodeling processes, such as sperm decondensation at fertilization [1]. In *Xenopus laevis* eggs, where it was first isolated, this highly acidic protein is thought to be in charge of nucleosomal core histones H2A/H2B storage. Upon fertilization, NP decondenses the densely packed sperm chromatin by means of extracting its specific basic proteins and replacing them with H2A/H2B, therefore enabling the assembly of somatic-type nucleosomes. NP is additionally involved in chromatin remodeling during early development, in particular it is required in the replication licensing mechanism, probably to extract linker-type histones from somatic chromatin, and can facilitate pluripotent cell reprogramming. NP (also designated NPM2) belongs to the nucleophosmin/nucleoplasmin family of histone chaperones [2]. Whereas NP roles have been particularly related to fertilization and embryogenesis, nucleophosmin (or NPM1) is ubiquitously and abundantly expressed in adult cells. It is enriched in the nucleolus, and serves multiple functions that affect cell growth and apoptosis, therefore disregulation of NPM1 is linked to several human cancers. Particular mutations of NPM1, that destabilize its structure, and cause its mislocalization to the cytoplasm, trigger acute myeloid leukaemia (AML). Apart from nucleophosmin and NP, the family includes the less known NPM3 and an invertebrate NPM-like.

NP is a homopentameric protein, composed of 200 residues, each subunit being built of two domains, namely core and tail. The core domain, corresponding to the N-terminal 120 residues, adopts an eight-stranded β-barrel structure, and is responsible for oligomerization, forming a ring with pentameric symmetry of 60 Å (diameter) and 40 Å (height) [3]. This compact core, shared by all NPM family members, confers an extreme stability to NP (see below). Probably, this is mainly due to a conserved network of hydrophobic interactions between the subunits, which, acting as a belt, firmly secures the pentamer. The C-terminal tail domain, instead, is conformationally flexible [4,5], therefore NP is considered "partially disordered" [6]. The tail harbors a segment rich in acidic residues (20 Asp and Glu within residues 120-150) termed "polyGlu", probably involved in histone binding , and a nuclear localization signal (NLS) that directs NP import into the nucleus.

The function of NP is activated through phosphorylation of up to 7 - 10 residues per monomer. NP phosphorylation degree correlates with *Xenopus* egg maturation, so that at the time of fertilization the protein is heavily phosphorylated and displays a maximal chromatin decondensing activity [5,7]. We have identified by mass spectrometry eight phosphorylation sites in natural NP: these phosphoresidues accumulate in flexible regions and loops, along both the core and tail domains, and cluster on a particular pole of the protein, known as distal face [8]. Phosphorylation causes a significant destabilization of the protein and we have made use of calorimetry (differential scanning calorimetry (DSC)) to dissect this effect, and its correlation with NP activation mechanism [9].

To fulfil its chromatin remodeling role, NP has to bind histones, basic proteins needed for packing of DNA. It acts as a reservoir for nucleosomal histones H2A/H2B, and is able to extract sperm specific basic proteins as well as linker-type histones, such as H1 from chromatin. The NP-mediated exchange of these more basic proteins with H2A/H2B results in a looser condensation state of chromatin [1,2]. We have thermodynamically described NP recognition of H2A/H2B and H5, a linker-type histone, by isothermal titration calorimetry (ITC) [10].

NP is the most abundant nuclear protein of *Xenopus* oocytes. Its nuclear import is mediated by the importin α/β heterodimer; in fact, NP is the prototypical substrate of this "classical" pathway, which is in charge of transport of most nuclear proteins [11]. Importin α recognizes a nuclear localization signal (NLS) in its substrates, which consists of a sequence segment with conserved basic residues, and itself associates to importin β [11,12]. The complex formed by importin α/β bound to the NLS cargo, traverses the nuclear envelope through the nuclear pore complexes. The transport relies on a gradient of the small GTPase Ran for directionality. The GTP-bound state of Ran is mostly nuclear and promotes the disassembly of the import complex once it reaches the nucleus, whereas in the cytoplasm, in the presence of Ran-GDP, the import complex formation is favoured [11,12].

Both importin α and β, belong to the karyopherin family of transport receptors, and their structures are constituted by a series of helical repeats, called ARM in the case of importin α and HEAT in importin β, that generate curved, flexible surfaces to bind their ligands. Importin α displays additionally a short N-terminal region for importin β binding (IBB domain) [12]. Most studies on the molecular basis of NLS recognition by nuclear transport receptors are so far limited to isolated domains of the proteins involved (e.g. using peptides corresponding to the NLS of NP and IBB of importin α) [13,14]. We have approached the thermodynamical characterization of the assembly of the complete complexes made of the full length proteins and have additionally built structural models of those import complexes [15]. NP loaded with histones can additionally incorporate importin α, generating large assemblies that could represent putative NP/histones co-transport complexes [16].

2. Calorimetry: Protein folding/unfolding and binding energetics

Calorimetry (DSC and ITC) is the most precise tool in the study of energetics of thermally-induced conformational transitions of proteins and their assembly with other molecules, small ligands or macromolecules. Extensive reviews have been published on the basic thermodynamic formalism, calorimeters' design and application of DSC [17-20] and ITC [18,20-24]. Moreover, surveys on ITC application are published annually since 2002 [25-28]. Calorimetry on proteins in general will be briefly summarized here and examples for NP will be thoroughly reviewed.

The excess heat capacity of a protein in solution, as a function of temperature, and the heat released or absorbed upon binding interactions are the quantities registered in the DSC and ITC experiments. Both the folding and binding events are described by the Gibbs free energy (ΔG), which determines the stability of the protein and the strength of association of molecules, respectively. The partitioning of ΔG into enthalpic (ΔH) and entropic ($T\Delta S$) terms is given by the basic thermodynamic equation:

$$\Delta G = \Delta H - T\Delta S \tag{1}$$

From the experimentally observed calorimetric curve, the DSC thermogram (typically an endothermic peak) and the ITC binding isotherm (exotherm or endotherm), a complete set of thermodynamic parameters of the studied folding/unfolding and binding phenomena is provided.

In DSC the values of the thermodynamic parameters of the folded-unfolded state equilibrium: transition midpoint temperature T_m (the temperature at the maximum of the excess heat capacity curve), the enthalpy of unfolding ΔH, calculated by the integral of the excess heat capacity function:

$$\Delta H = \int_{T_o}^{T_u} c_p dT \tag{2}$$

where T_o and T_u are the temperatures of the onset and completion of the transition, respectively, and c_P (the heat capacity change associated with unfolding) can be determined in a model-independent way [29]. In addition, the width at half-height of the transition $T_{m1/2}$ is a measure of the cooperativity of the transition from folded to unfolded state.

In ITC, the binding affinity K_b ($K_b = e^{-\Delta G/RT}$, R is the gas constant and T is the absolute temperature), the enthalpy change ΔH and the stoichiometry N of the binding interactions are determined by fitting the experimentally obtained binding isotherms assuming a model that well describes the binding process. While one binding constant describes a simple 1:1 molecular interaction, complex macromolecular recognition processes are described by model-independent macroscopic and model-dependent microscopic association constants, that account for the overall binding behaviour and for the association at each binding site, respectively [30,31]. Model independent analysis of more complex binding data, with two or more binding sites, based on general binding polynomial formalism, developed by Freire et al. [31], allows the type, independent or cooperative, of the binding interactions to be assessed. Methodology and analysis for heterotropic ligand binding cooperativy, i.e. for two or more different ligands binding to one protein has also been elaborated by Velázquez-Campoy et al. [32,33].

ITC is a suitable technique for characterizing allosteric interactions and conformational changes in proteins [32,34-37].

ITC also allows determination of the heat capacity change of binding interactions, Δc_P, from the temperature dependence of the enthalpy change:

$$\Delta c_p = \partial \Delta H / \partial T \tag{3}$$

with the assumption that the apparent heat capacities of the free molecules and the complex are constant over the temperature range under study. The changes in the heat capacity associated with protein-protein binding originate mostly from changes in the hydration heat capacity due to burial of polar and nonpolar groups upon complex formation and the loss of conformational degree of freedom upon binding [38-40]. Hence, Δc_P can be calculated in terms of the change in the accessible surface areas (apolar (ASA_{ap}) and polar (ASA_{pol})) upon the formation of protein-protein complex using the semi-empirical relationship [39,41-42]:

$$\Delta c_p = 0.45 \Delta ASA_{ap} - 0.26 \Delta ASA_{pol} \quad cal\ K^{-1}mol^{-1} \tag{4}$$

A good correlation between the experimentally determined and the calculated from structural data Δc_P values has been found in some cases [43-45], however significant difference was reported in other cases [42,46], suggested to be a consequence of changes in the conformational states and significant dynamic restriction of vibrational modes at the surface of the complex, as well as folding transitions coupled to the association event.

The values of Δc_P can also be used to estimate the entropic component due to desolvation of the surfaces of both interacting proteins buried within the binding interface:

$$\Delta S_{solv} = \Delta c_p \ln \left(T / T^* \right) \tag{5}$$

where $T^* = 385.15$ K is the temperature of entropy convergence [47,48] and to further decompose the entropic term, which besides the solvation term contains two more contributions (conformational, ΔS_{conf}, associated with changes in conformational degree of freedom and rotational-translational ΔS_{rot-tr}, (ΔS_{rot-tr} = -7.96 cal mol^{-1}K^{-1} [49] which accounts for changes in rotational/translational degrees of freedom):

$$\Delta S = \Delta S_{solv} + \Delta S_{conf} + \Delta S_{rot-tr} \tag{6}$$

Additional information on protonation/deprotonation effects coupled to the binding interactions can be provided by titration experiments in various buffers of different ionization enthalpy, ΔH_{ion}. The number of protons, n_{H^+}, exchanged between the macromolecular complex and the bulk solution, and the binding enthalpy, ΔH_{bind}, can be calculated from the dependence of the calorimetrically observed enthalpy change, ΔH_{obs}, and ΔH_{ion} [42,50,51]:

$$\Delta H_{obs} = \Delta H_{bind} + n_{H^+} \Delta H_{ion} \tag{7}$$

To decompose the free energy of binding into electrostatic and non-electrostatic contributions one has to study the ionic strength effect on the binding thermodynamics and analyse the data according to the Debye-Hückel approximation [52].

Valuable information on the hydration or solvent exposure of a polypeptide can be obtained by the absolute heat capacity [29]. A DSC method to accurately determine the absolute heat capacity of a protein from a series of calorimetric thermograms obtained at different protein concentrations has been described in [53]. The slope of a plot of the excess heat capacity versus the protein mass in the calorimetric cell is related to the absolute C_p:

$$m = C_p - \upsilon_p \tag{8}$$

where m is the slope of the linear regression of the plot and υ_p is the partial specific volume of the protein. This information can be related to the integrity of the native state or the presence of residual structure in the denatured state.

The calorimetric transitions in many cases are irreversible and scanning rate dependent, suggesting that the denaturation process is kinetically controlled [54-56]. Appropriate kinetic models were applied to analyse the irreversible unfolding process, after obtaining a set of thermograms at various scanning rates. An irreversible protein denaturation event can be described in some cases by a simplified "two-state irreversible" kinetic model [54,57], assuming that only the native/folded and denatured/unfolded states are significantly populated during the denaturation. Mathematical expressions were derived to calculate the activation energy, E_a, of the denaturation transition [54-58], using diverse experimental information from the calorimetric transition:

i.   the values of the rate constant of the transition, k, at a given temperature:

$$k = A \exp(-E_a / RT) \tag{9}$$

where E_a is the activation energy and A is the frequency factor.

The rate constant of the reaction at a given temperature T is given by:

$$k = \nu\, c_p /(Q_t - Q) \tag{10}$$

where v is the scanning rate (K/min), c_P the excess heat capacity at a given temperature, Q is the heat evolved at that temperature and Q_t the total heat of the calorimetric transition.

ii.   the dependence of the heat capacity evolved with temperature expressed as:

$$\ln\left[\ln Q_t /\left(Q_t - Q\right)\right] = E_a / R \left(1/T_m - 1/T\right) \tag{11}$$

iii.   the heat capacity c_P^m at the transition temperature T_m, where the activation energy can also be calculated by the following eq.:

$$E_a = eRT_m^2 c_P^m / Q_t \tag{12}$$

This "two-state" kinetic model has described the unfolding of bacteriorhodopsin [58], rhodopsin [59], plastocyanin [60], the major light harvesting complex of photosystem II [61], nucleoplasmin (see below, [9]) and some other proteins [62,63]. This model however cannot describe all cases of irreversible protein denaturation [64]. On the other hand, Davoodi et al. [65] have shown that scanning-rate dependence of DSC thermograms is not limited to irreversible processes only.

DSC can also be used to indirectly study ligand binding to proteins and for analysis of very tight binding that can not be analysed by ITC or other spectroscopic methods [66]. In addition, more comprehensive description of the binding energetics can be derived combining the two techniques, ITC and DSC [18].

Recently DSC was also recognized as a novel tool for disease diagnosis and monitoring [67-70]. Calorimetric studies of blood plasma/serum have revealed a typical DSC thermogram for healthy individuals, whereas pronounced changes in thermograms for diseased subjects, including oncopatients, have been reported. Validation of the technique as an efficient tool for disease diagnostics needs further investigations of a large number of diseases.

Besides the classical application of ITC in studies of binding interactions, it has been proven as a powerful technique in diverse fields like drug discovery and lead optimization, nanotechnology, enzyme kinetics, etc. [71,73].

Kinetics of ligand binding to RNA and the subsequent RNA folding have recently been characterized by the so called kinITC [74]. ITC has also permitted documentation of the energy landscape of tertiary interactions along the RNA folding pathway [75]. Thermodynamic parameters, ΔH and Δc_P, of rigid amyloid fibril formation from monomeric β-microglobulin, associated with degenerative disorders have also been determined by ITC [76]. Recently a protocol for novel application of the technique has been elaborated, in which ITC is used as a tracking tool, combined with chromatography, for identification of target protein in biomolecular mixture [77] and it has been suggested to be valuable when the target protein or ligand is unknown. References for the wide spectrum and examples of novel applications of ITC can be found in the surveys published each year in the Journal of Molecular Recognition [25-28].

3. Nucleoplasmin thermal stability. Differential scanning calorimetry

NP is remarkably stable against chemical and physical challenges, including heat; e.g. the T_m of recombinant, non phosphorylated protein is 110.1°C [4,9]. This extreme stability, which is related, as previously mentioned, to the structural scaffolding role of the core domain, is *per se* an attractive issue to be thermodynamically described. We have characterized by DSC and other techniques the stability properties of NP and how they are related to the functionality of the protein. It should be mentioned that the overpressure used in the calorimeter allows to asses melting points above water boiling temperature, by contrast to other spectroscopic techniques.

We have shown that the stability of NP is solely due to the core domain, the T_m of the isolated core (117.6°C) being still higher than that of the full length protein [9]. The slight destabilizing influence of the tail is explained by its strong acidic character, with negatively charged clusters, such as "polyGlu"; electrostatic repulsion is expected to occur between tails and the also negatively charged core domain. This is reflected by the fact that the full length protein is most stable at pH close to its theoretical isoelectric point (pI=5.1), when its stability equals that of the core domain [9].

Analysis of the chemical denaturation of NP by fluorimetric and biochemical techniques has allowed to describe the unfolding mechanism of NP, in terms of a two-state process, where the pentamer dissociation is coupled with unfolding of the monomers, with no evidence of (partially) folded subunits ($N_5 \leftrightarrow 5U$) [78]. Both chemical and thermal unfolding of NP are reversible processes, while denaturation of the isolated core domain is reversible if chemically induced but irreversible upon heating [9,78]. This different behaviour suggests that the charged tail domains favour the solubility of the full length protein after thermal unfolding.

4. Effect of NP activation. Interplay between function and stability

NP activation, mediated by phosphorylation of multiple residues, implies an energetic cost for the protein. We have observed that NP extracted from *Xenopus* oocytes, corresponding to an intermediate phosphorylation state, is significantly destabilized with respect to recombinant, non phosphorylated NP ($T_m \sim 94.4°C$, $\Delta H \sim 80$ kcal/mol) [4]. Egg NP, which represents the most active protein in the final stage of egg maturation, exhibits a further destabilization ($T_m \sim 75°C$, $\Delta H \sim 50$ kcal/mol) [9]. This correlation between phosphorylation degree and loss of stability has been also characterized by chemical unfolding experiments [78].

In order to explore the conformational consequences of NP activation, we assessed the impact of phosphorylation in particular sites on the protein stability. Apart from the experimental evidences pointing to CKII and mitosis promoting factor (MPF) as probable kinases that modify NP [7,79], the amount and identity of kinases phosphorylating NP has not been elucidated. On the other hand, NP can be phosphorylated *in vitro* only with very low yield. As an alternative approach to obtain homogeneous preparations of active NP with a defined modification level, we designed a series of phosphorylation mimicking mutants, in which different Ser and/or Thr residues representing phosphorylatable residues were substituted for Asp [8,9,80]. Most mutation sites correspond to phosphoresidues identified by mass spectrometry analysis of egg NP [9]. However, taking into account that some phosphoresidues might have remained undetected by the proteomic analysis, due to incomplete sequence coverage and/or heterogeneity of the NP natural samples, additional residues were mutated on the basis of prediction software, N-terminal amino acid analysis, sequence comparison within the NP family and structural considerations [8,80].

The mutation sites, which are indicated in Figure 1, can be classified in three groups: 1) mutations in the flexible, N-terminal segment of the protein, not visible in the 3D structure of the core domain (residues 2, 3, 5, 7, 8), 2) mutations of residues located in loop regions of the core domain distal face (15, 66 and 96) and 3) mutations in the tail domain (residues 159, 176, 177, 181 and 183). Apart from the group of three residues within the structured core domain, at least group 1 is expected to face also the distal pole of the protein, which is most probably implicated in histone binding [10,80]. A collection of NP mutants (full length and core domains) were generated, combining the three groups of mutations, as indicated in Figure 1.

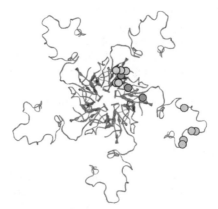

Figure 1. Activation of recombinant NP achieved through phosphomimicking mutations. Their location is highlighted on our model of full length NP based on the crystal structure of the core domain [3] and SAXS data [10]. Orange circles denote substitutions of residues 2, 3, 5, 7, 9 for Asp in the N-terminal segment ("group 1"); red ones correspond to substitutions at 15, 66 and 96 ("group 2"), and green ones to mutations in residues 159, 176, 177, 181, 183 of the tail domain ("group 3"). NP5D carries only mutations in the tail; NP8D harbors groups 1 and 2; NP10D groups 1 and 3; and NP13D comprises all mutations. For the sake of clarity, the positions are shown in only one monomer

By contrast to recombinant, non-phosphorylated NP, which shows negligible ability to decondense chromatin, phosphorylation mimicking mutations render the protein active to varying extents depending on the number and position of mutations. The mutants are capable of decondensing *Xenopus* demembranated sperm nuclei and extracting sperm-specific basic proteins, as well as linker-type histones from chromatin. The core domain isolated from natural, hyperphosphorylated, egg NP is (partly) active in decondensing chromatin, and a recombinant core domain with 8 substitutions (CORE8D, with groups 1 and 2) resembles these functional properties [80]. Nevertheless, full activity can only be attained through accumulation of negative charges (or phosphoresidues) along both the core and tail domains of NP: the mutant NP13D reproduces the functionality of egg NP [8].

Comparison of the thermal unfolding profiles of wild type and mutant core domains reveals that the activating mutations strongly decrease the thermal stability of the protein (Figure 2). Destabilization is probably due to the electrostatic repulsion in the oligomer (already negatively charged at neutral pH), which becomes more intense in the mutants (see inset in Figure 2).

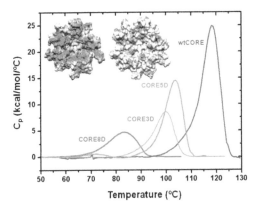

Figure 2. Phosphorylation mimicking mutations decrease the thermal stability of NP core domain. The effect of the mutations on the charge of the protein is also shown, by comparing the surface of the crystal structure of the mutant CORE8D [9] (left) and wild type CORE [3] (right), viewed from the distal face, and colored according to the electrostatic potential

The mutant CORE3D (with group 2 mutations) is less stable, in spite of harbouring fewer substitutions than CORE5D (group 1 mutations), which could be due to the fact that the three residues 15, 66 and 96 locate in structured regions of the protein, whereas the five N-terminal mutations are in a flexible segment that could be re-arranged to alleviate the electrostatic repulsion in NP. The combination of both groups of mutations makes CORE8D the most unstable mutant core, as expected. The strong destabilizing effect (e.g. ΔT_m of 34.5°C in the case of CORE8D) suggests a conformational change in the protein. Phosphorylation does not induce, however, significant changes in the secondary structure of NP [5,9]. Furthermore, we have solved the crystal structure of CORE8D and found that surprisingly enough it is almost identical to that of wild type core domain [9] (see Figure 2).

On the other hand, the activating mutations seem to affect the dynamics of the core domain [9]. The irreversible thermal unfolding of the core can be described as a scanning rate-dependent transition between two states, native and irreversibly denatured. From different mathematical expressions making use of diverse parameters from the calorimetric transition (eqs. 9-12), the activation energy E_a of the denaturation was calculated and compared for wild type CORE and CORE8D. We obtained a higher E_a value for wtCORE (69.8 ± 2.6 kcal/mol) than for CORE8D (52.1 ± 2.4 kcal/mol), indicating that the mutations destabilize also kinetically the core domain, reducing the energy barrier of the transition to the unfolded state [9]. In addition, to further characterize the conformational change associated with protein activation, the excess heat capacity c_P of wtCORE and CORE8D was measured at various protein concentrations (at 37°C), in order to calculate the absolute heat capacity, C_p, of their native states, which is related to solvent exposure of protein hydrophobic groups (eq. 8, see above). The obtained C_p values were 0.23 and 0.42 cal K^{-1} g^{-1} for wtCORE and CORE8D, respectively, suggesting faster dynamics or faster conformational fluctuations in the mutant [9]. Furthermore, the activation process affects the hydrodynamic properties of the protein (see below).

To understand the contribution of both NP domains to its activation mechanism, we also characterized the function and stability of full length NP, with the three groups of mutations and combinations thereof (Figure 3). Substitutions located in the core domain (NP8D) affect the protein stability to a greater extent than those in the tail domain (NP5D), highlighting again that addition of charges in structurally well defined locations is more deleterious for the stability of the protein than in flexible regions. The most active mutant, NP13D, is also the most unstable (T_m ~ 55.2°C, ΔH ~ 17.9 kcal/mol). Taking into account that at neutral pH aspartic acid has one negative charge, while a phosphoryl group would display an average negative charge of -1.5 [81], 13 Asp would be a reasonable approximation of 7-10 phosphates per monomer in egg NP; however, the fact that this mutant is less stable than egg NP reflects that the conformational properties of phosphorylated NP may not be exactly reproduced [9].

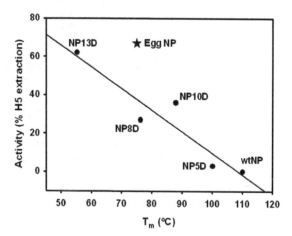

Figure 3. Inverse correlation between NP activity (expressed as percentage of histone H5 extracted from chicken erythrocyte chromatin, in a solubilization assay [8]) and thermal stability (T_m as measured by DSC). Linear regression of the phosphomimicking mutants data is shown

The mutations-induced destabilization of NP is also readily observed by chemical unfolding experiments [9]. Since NP denaturation proceeds through pentamer dissociation intimately coupled to unfolding of the monomers, the activation mechanism, in spite of not affecting conformationally the protein at the level of secondary structure and tertiary structure of the core domain, seems to weaken its quaternary interactions. In support of this notion, we have observed, by size exclusion chromatography and dynamic light scattering, that the activating mutations induce an expansion of the NP pentamer dimensions in solution, both in the core domains (from an average diameter of 64.5 Å to 68.8 Å for CORE8D, as measured by DLS) and the full length mutants (from 93.7 to 99.5 Å in the case of NP13D) [9]. Considering the similarity between the crystal structures of inactive, wild type, and active, mutant core domain, this "swelling" must affect mainly flexible regions of the protein, such as the N-terminal segment, loops of the core domain, and the tail domain.

Therefore, in NP, an inverse correlation exists between activity and stability (Figure 3), the higher the histone chaperone activity performed by NP, the lower its thermal and chemical stability, and larger its dimensions in solution.

In summary, we observed that NP activation mechanism, that depends on the accumulation of negative charge, probably on flexible regions of the distal pole of the protein, implies a destabilizing cost and an expansion of the oligomer in solution. The destabilizing mechanism seems to be the electrostatic repulsion in the pentamer, that weakens the quaternary interactions (tending to "open" apart the structure), which are essential for the stability of this protein. However, the loss of stability does not compromise, under physiological conditions, NP function or folding, which is granted by the extremely stable core domain. Moreover, the activation penalty may explain why this protein, from a mesophilic organism, displays such a remarkable thermal stability: it is necessary to afford the strong destabilization upon activation.

5. Nucleoplasmin chaperoning function studied by isothermal titration calorimetry

The high number of positive charges that histone proteins carry makes them prone to unproductive interactions with nucleic acids and other cellular components. Therefore free histones eventually do not exist within the cellular context and need to be escorted by histone chaperones, which shield their charge, and facilitate their controlled transfer during nucleosome assembly or reorganization. To perform its function, nucleoplasmin has to bind both linker-type and nucleosomal histones. Thermodynamics provided a detailed knowledge of NP-histone complex formation and elicited how NP carries its chaperoning activity [10].

The experimental isotherms of the binding interactions of NP with histones, H5 and H2A/H2B, and the enthalpic and entropic contributions to the Gibbs free energy for the first binding site are summarized in Figure 4.

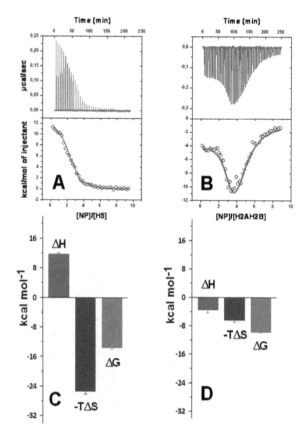

Figure 4. Binding data of NP interactions with the linker, H5, and nucleosomal core, H2A/H2B, histones. (A, B) Baseline-corrected instrumental response of NP titration with successive additions of H5 and H2A/H2B (upper panels); integrated data and the fits of the binding isotherms (solid lines) according to a negative cooperativity model (see text) for H5 and H2A/H2B (lower panels). (C, D) Thermodynamic parameters (ΔG, ΔH, -TΔS) of the assembly of the first histone, H5 and H2A/H2B, molecule with NP

ITC data reveal that NP can accommodate five histone molecules utilizing a negative cooperative binding mechanism with dramatic difference in the binding strength. The binding affinity of histones for the first site is moderate for nucleosomal core ($\Delta G = -9.8\pm0.1$ kcal/mol, $K_b=1.5\times10^7$ M^{-1}) and extremely high for linker ($\Delta G = -13.6\pm0.4$ kcal/mol, $K_b=10^{10}$ M^{-1}) histones (Figure 4, C and D), which can provide the basis for its histone exchange capabilities. The binding isotherms of the complex formation of histones with NP were analyzed using a site specific cooperative binding model. The model, developed especially for NP-histone interactions, considers negative cooperative interactions for both adjacently and non-adjacently bound histones and fit the experimental data better than an independent binding sites model and a general model based on the overall association constants. Eight thermodynamic parameters: four association constants (intrinsic association constant (K) and cooperativity binding parameters: k_1 (associated with the binding of an additional ligand), k_2 (binding of a ligand with contact to one nearest-neighbour) and k_3 (binding of a ligand with contact to two nearest-neighbours)) and four enthalpies were defined in the cooperativity model (details on the model can be found in [10] and in Supplementary Material of [10]).

The binding of histone molecules upon occupancy of the first binding site progresses with an energetic penalty, with exception of H2A/H2B molecules that bind to a non-adjacent site ($k_1 = 1$). Therefore, negative cooperativity was observed for all four additional H5 molecules ($k_1 < 1$), while only for H2A/H2B histone dimers bound to NP adjacently. Since the source of the cooperativity interaction may be an allosteric conformational change in NP induced by histone binding or a direct histone-histone interaction upon binding, or/as well as a combination of both, our results indicate different origin of the negative cooperativity for the binding of H5 and H2A/H2B to NP. Hence the main source of the cooperative binding interaction of H2A/H2B dimers is H2A/H2B-H2A/H2B interaction, whereas a conformational change in the NP pentamer upon binding of the first H5 molecule should provide a less favourable binding interface for the next histone molecules through energetic communication.

Somewhat surprising, considering the strong opposite charge of NP and histones, the binding of both histone types to NP is dominated by a favourable entropic term indicating a strong contribution of the hydrophobic effect to the binding affinity (Figure 4, C and D). The enthalpic term also contributes favourably to the binding energy of H2A/H2B, while unfavourable enthalpy changes counterbalance the entropic contribution to the free energy of H5 binding (Figure 4, C and D).

Furthemore, and contrary to the generally accepted major determinant of tail "polyGlu" tract in histone binding, the thermodynamic analysis as well as the low resolution structural models of NP/histone complexes, constructed by small angle X-ray scattering (SAXS) [10], demonstrate clearly that both NP domains are involved in the interaction with histones.

This was evidenced by comparing the binding energetics of the full-length protein with that of isolated core domain (CORE). Interestingly, NP core domain contributes equally to the intrinsic binding energy of H5 and H2A/H2B ($\Delta G = -8.2$ kcal/mol). The tail domain of NP provides an additional thermodynamic driving force (estimated as the difference between the binding free energies of histones to NP and CORE, $\Delta\Delta G_{NP\text{-}CORE}$) (Figure 4 and Figure 5) for the much stronger binding of H5 ($\Delta\Delta G_{NP\text{-}CORE}^{H5} = -5.5$ kcal/mol) compared to H2A/H2B ($\Delta\Delta G_{NP\text{-}CORE}^{H2AAH2B} = -1.6$ kcal/mol) suggesting that this domain is particularly essential in the binding to H5 molecules.

To approach an activity/energetics relationship, we analysed the energetics of histone association with NP variants with phosphorylation-mimicking mutations in both the core and tail domains (NP8D, NP13D, CORE, CORE8D).

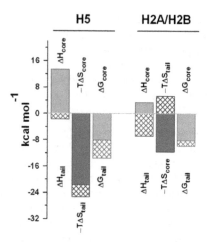

Figure 5. NP core (full bars) and tail (crossed bars) domain contributions to the intrinsic ΔG of their binding to linker and nucleosomal histones

As mentioned above the NP activity is regulated by its phosphorylation state. Insertion of mutations (8 and 13) gradually enhanced the binding affinity and affected to different extent the changes in the Gibbs energy contributors, the entropic and enthalpic terms. This reflects a strong impact of phosphorylation mimicking mutations in both core and tail domains of NP on its recognition by histones (Figure 6). The strongest affinity observed for the NP variant with the highest number of mutations, NP13D, is compatible with the fact that it mimicks the activity of the hyperphosphorylated native protein and can explain the protein activation through post-translational modifications.

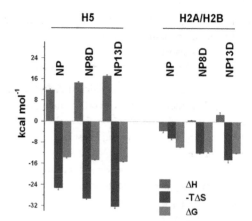

Figure 6. Effect of phosphorylation mimicking mutations on the binding energetics. Bar graphs comparing the intrinsic Gibbs energy, enthalpy and entropy changes, for the intrinsic binding of the two histone types, H5 and H2A/H2B, to NP and the phosphomimicking mutants NP8D and NP13D

Although the hydrophobic interactions are the major source of NP/histone binding free energy (about 80% of the intrinsic free energy for H2A/H2B and about 60% of that for H5), electrostatic and polar interactions between the acidic NP and basic histones also play an important role, either in direct binding or helping in orienting properly the binding partners, given the structural features of NP and histones. In order to get more insight into the nature of binding interactions we studied the ionic strength effect on the binding energetics (Figure 7, left panel). We found that despite the highly charged nature of H5 and NP, the non-electrostatic interactions contribute stronger to the stabilization of NP/histone complexes than the electrostatic ones (Figure 7, right panel). The significantly lower observed free energy of binding ΔG for H2A/H2B compared to H5 originates from lower ΔG_{el} (electrostatic) term (the non-electrostatic term ΔG_{nel} is comparable for H2A/H2B and H5), that should reflect the distinct number of positively charged residues in each histone type. For H2A/H2B and H5 binding to the NP13D mutant ΔG is -9.8 ± 0.1 and -11.9 ± 0.14, and the ΔG_{el} term -2.6 kcal/mol and -5.1 kcal/mol, respectively (Figure 7, right panel).

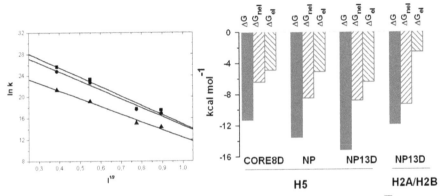

Figure 7. Ionic strength dependence of the association constant of H5 binding to the first (■), non-adjacent (•) and adjacent (▲) binding sites of NP13D variant (left panel). Extrapolation of $\partial\ln(K) / \partial I^{1/2}$ to 1M NaCl yields the non-electrostatic contribution ΔG_{nel} to the binding energy ΔG. Contribution of ΔG_{nel} and the electrostatic contribution, ΔG_{el}, to ΔG for H5 binding to CORE8D, NP and NP13D, and of H2A/H2B to NP13D (right panel)

ITC data also show that the NP flexible tail domain undergoes a histone binding-induced transition to a more structured or ordered state. This follows from the conformational entropy difference between full length proteins and core domains. We estimated from the heat capacity change, that there is a conformational entropy loss of ca. -20 kcal/mol upon H5 binding to the full-length protein as compared to the core domain (and even higher ca. -33 kcal/mol for H5 binding to the mutant proteins, NP8D and CORE8D), that can be attributed to the ordering of the intrinsically disordered nucleoplasmin tails [5] when bound to histones and indicates that NP tails do establish contacts with the histone molecules.

On the other hand, Δc_P (obtained from the temperature dependence of ΔH, presented for the NP13D variant in Figure 8) is smaller for the core domain NP variants compared to the full-length NP that would indicate a smaller molecular surface area involved in the binding of H5 to the core domain fragments than to full length NP.

Since no high resolution structural data are available for the NP/histone complexes the experimental Δc_P heat capacity changes cannot be compared with the ones estimated from structural data. We therefore roughly estimated the area buried within the binding interface from the SAXS data, in terms of "dummy" atoms of the corresponding "phase" that are in contact with the atoms of another "phase" in MONSA models (for details on SAXS experiments and data analysis see ref.[10]).

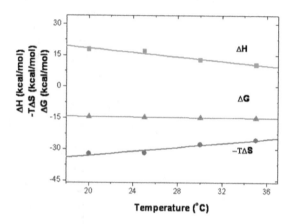

Figure 8. Temperature dependence of the thermodynamic parameters of the binding of the first H5 molecule to NP13D mutant. Intrinsic enthalpy (ΔH, ■), entropy (-TΔS, •) and free energy (ΔG, ▲) of binding. The heat capacity change Δc_P ($\Delta c_P = (\partial \Delta H / \partial T)$) is determined from linear regression analysis of ΔH data (solid line). The intrinsic free energy of binding is almost independent of temperature reflecting compensation of the enthalpic and entropic terms

The interaction interface area corresponding to NP tail/H5 is approximately double that of NP core/H5, whereas NP tail/H2A/H2B is half of that of NP core/H2A/H2B, which reflects strong difference in the binding of NP tails to both histone types. Although the ratio of the interaction interface areas NP core/NP tails is a rough estimate, it well compares with the ratio estimated from the experimentally determined heat capacity changes, $\Delta c_P^{NPcore}/\Delta c_P^{NPtails}$ ($\Delta c_P^{NPtails} = \Delta c_P^{NP} - \Delta c_P^{core}$).

The significant differences between the intrinsic association constants and the cooperative character of NP binding to the nucleosomal and the linker histones defines different "affinity windows" for NP binding from picomolar to nanomolar and from nanomolar to micromolar for H5 and H2A/H2B, respectively. This difference in recognition of nucleosomal and linker histones might provide an efficient mechanism for regulation of the dynamic histone exchange and might allow NP to fulfill its histone chaperone role, simultaneously acting as a reservoir for the core histones and a chromatin decondensing factor. Our data are compatible with the traditional model where NP facilitates nucleosome assembly by removing the linker histones and depositing H2A/H2B dimers onto DNA [1,2,82].

These data provided new insight into NP/histones assembly and interactions. It should be emphasized that the binding affinity of NP is enhanced upon insertion of phosphorylation-mimicking mutations that explains the protein activation through post-translational modifications. Importantly the data reveal a negative cooperativity-based regulatory mechanism for the linker histone/nucleosomal histone exchange, that in general renders proteins operative in a wider concentration range [83], with significantly populated intermediate liganded states. The employed site-specific cooperativity model, an extension of a previous one that analyses the interaction of another pentameric protein (cholera toxin, with the oligosaccharide portion of its cell surface receptor) considering only nearest-neighbour cooperative interactions [84], has potential application in studies of other macromolecular complexes between proteins sharing structural complexity with NP and their ligands.

6. Recognition of nucleoplasmin and histones by nuclear transport receptors

Nucleoplasmin, that possesses a classical bipartite NLS targeting sequence in each tail domain, is a prototypic substrate of the best characterized route for protein import into the nucleus, which is mediated by importin α/β heterodimer. However, as mentioned in Introduction, most structural and energetic approaches on cargo-import receptor recognition have been achieved merely using peptides carrying the corresponding NLS or IBB sequence [13, 85-91], and to date only two studies (besides ours [10]) deal with assembly of a macromolecular transport complex of full-length proteins [92,93]. On the other hand there has been paid little attention to nuclear import of oligomeric proteins. Therefore understanding the molecular basis of recognition of an oligomeric cargo as nucleoplasmin by its transport receptors, importin α/β heterodimer, would shed light on the arrangement of a large macromolecular nuclear import complex.

We obtained saturated NP/importin α/β complexes proving that all five available NLS binding sites of NP can be occupied by importins. Whereas *in vivo* binding of one α/β heterodimer to any protein should be enough to deliver it to the nucleus, it has been reported that the presence of multiple NLSs in NP [94] enhances its nuclear accumulation, suggesting that the number of NLSs might govern the traffic rate, which would play an advantage for oligomeric nuclear proteins, provided with multiple recognition sites. The binding isotherms of the NP/α/β complex formation (Figure 9A) were well fitted with an independent binding sites model, reflecting that NP makes use of different energetic scheme for assembling with histones and importins, most likely due to the involvement of dissimilar binding surfaces. The binding reaction is enthalpy-driven and counterbalanced by an unfavorable entropy change (Figure 9B) resulting in a relatively high-affinity interaction, K_b = 18.5 × 10^6 M^{-1} (K_d = 57 ± 15 nM). The entropic penalty most probably reflects an ordering effect on the otherwise flexible and mobile NLS motifs [5] upon the interaction event. This loss of conformational flexibility of the NLS segment in the NP tails [5] is estimated to correspond to conformational entropy change ΔS_{conf} = -282 cal mol^{-1} K^{-1} (using eq. 6) that dominates the entropic penalty. This conformational entropy change is unfavorable and greater than the favorable solvation entropy (ΔS_{solv} = 223 cal mol^{-1}K^{-1}, calculated from eq. 5), associated with hydrophobic interactions, thus resulting in an unfavorable entropy contribution to the Gibbs free energy of binding (Figure 9C).

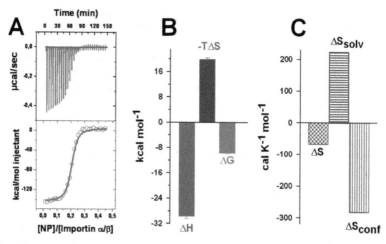

Figure 9. Energetics of NP assembly with the nuclear transport receptor importin α/β. Binding isotherms: the upper panel represents baseline-corrected instrumental response of importin α/β titration with NP; the lower panel shows the integrated data and the fit of the binding isotherm (solid line) by an independent binding site model (A). Enthalpic (ΔH) and entropic (-TΔS) contributions to the free energy (ΔG) of binding (B). Dissection of the binding entropy, ΔS, into solvation, ΔS_{solv}, and conformational, ΔS_{conf}, terms (C)

Similar binding mode and thermodynamic parameters (ΔG = -8.67 ± 0.1 kcal/mol, ΔH = -15 ± 0.3 kcal/mol and stoichiometry 5 per NPM pentamer) characterize the recognition of nucleophosmin (NPM), an abundant nucleolar protein, structurally homologous to nucleoplasmin (as mentioned in the Introduction) and related to oncogenic transformation, by the nuclear transport machinery (unpublished data), which supports an equivalent import mechanism for both chaperones.

Full-length importin α, not assembled with importin β, is also able to bind NP, albeit with a lower apparent affinity (K_d = 513 ± 87 nM) and with a lower enthalpic contribution to the free energy of binding. The loss of apparent affinity comes from the fact that the importin α N-terminal domain, the IBB domain, which contains a similar sequence to the NP-NLS, exerts an autoinhibitory role in the binding process [90,91] because in the absence of importin β, it occupies the NLS binding site and therefore it must be displaced by NP-NLSs. In this regard a truncated importin mutant lacking the IBB domain, ΔIBB-importin α or Δα, shows a similar affinity for NP (K_d = 54 ± 6 nM) as importin α/β [15].

Since protein phosphorylation is one of the mechanisms that up- or down-regulate nuclear transport [95,96] and it had been described that phosphorylated nucleoplasmin presents higher import rate than its unphosphorylated form [79], we studied how phosphorylation of NP affects its interaction with the import receptor. Nevertheless, phosphorylation mimicking mutations in residues close to NLS sequence, as in mutant NP13D, which shows high binding affinity to histones and is active in histone chaperoning, do not modulate the interaction with importin. NP13D mutant displays the same binding strength (ΔG), though different ΔH and $-T\Delta S$ terms, compared to unphosphorylated protein (Figure 10). No effect has been observed when phosphorylated monopartite NLS from simian virus 40-large T antigen interacts with importin α [97]. Altogether, phosphorylation-mediated regulation of nuclear import must involve interactions other than post-translationally modified NLS with α/β importin.

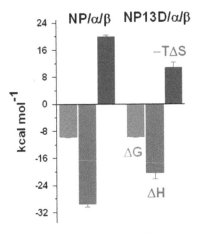

Figure 10. Comparison of the energetics (ΔG, ΔH and $-T\Delta S$) of importin α/β binding to NP and NP13D mutant

Importin binds similarly to a peptide corresponding to the NLS sequence when the latter is isolated or in the context of the full-length NP macromolecule, suggesting that no other regions of NP contribute significantly to the binding. The similar large negative Δc_P values, -817 and -796 cal mol^{-1} K^{-1} for NLS and full-length NP respectively, also suggest that the surface area buried within the binding interface is comparable in both cases. It is not surprising that the NLS recognition is not significantly affected by the protein context considering the flexibility displayed by NP tail domains harboring the NLS segments. The same notion applies to the interaction of importin α IBB domain with importin β. The former domain binds likewise to importin β independently of whether it is an isolated peptide or connected to the ARM domain of importin α, as is evidenced by the good correspondence of the heat capacity Δc_P value of -727.4 cal mol^{-1} K^{-1}, predicted from the X-ray structure of importin β bound to the IBB domain of importin α [98], with buried polar and apolar solvent accessible areas (ΔASA) of 1402.5 and 2426.7 Å^2, respectively (eq. 4), and the experimentally determined Δc_P = -840 cal mol^{-1} K^{-1} value for full length α/β interaction. Moreover, both proteins behave as independent units when they form the heterodimer complex since they display the same thermal stability as the one they exhibit when free in solution, thanks to flexibility of the linker between the IBB and the rest of importin α [15]. These data support the idea that α residues which act as link between both importins exhibit such flexibility that allows each of the importin entities to interact with a wide range of ligands during the nuclear translocation process.

Given the fact that the NP/α/β complexes are formed by multiple proteins that present flexible domains, SAXS technique has provided valuable information about the structure of those assemblies. Multiple models of NP fit equally well the experimental SAXS data reflecting the inherent flexibility of the particle, due to the adaptable linkers between the NP core domain and the NLS (residues 121-154 of NP) [15], which allows the accommodation of five bulky α/β heterodimers per NP pentamer. This 3D in solution structural model, the first one for a complete nuclear transport complex with an oligomeric cargo, is consistent with the notion that the canonical binding elements (NLS and IBB) are the ones determining the molecular basis of the recognition. The multidomain NP/α/β complex remains stable by virtue of two attachment points: recognition of the NLS by importin α and recognition of the IBB domain by importin β, which otherwise allow for conformational flexibility. This modular and articulated architecture might facilitate the passage of such a large particle through the nuclear pore complex.

Due to their highly basic nature histones need nuclear import receptors to be transferred to the nucleus, and most of the pathways described are mainly mediated by karyopherins of β family [99,100]. On the other hand, histones present multiple NLS-like motives and are also recognized by importin α family members for nuclear targeting [101]. Accordingly, we observed that both nucleosomal and linker histones bind to importin β (unpublished data), as previously demonstrated, and to importin $\Delta\alpha$ [16]. The high affinity exothermic binding interactions (Figure 11) suggest specific recognition events of importin $\Delta\alpha$ by H5 and H2A/H2B. Regardless of the different stoichiometry, two importin $\Delta\alpha$ per H5 and one per H2A/H2B, the thermodynamic parameters are quite similar, the apparent binding affinity and the enthalpy are in the order of 9 and 28 nM, and −20 and −17 kcal/mol for H2A/H2B and H5, respectively. Similar binding energetics, though higher stoichiometry (five per NP pentamer), characterizes the assembly of importin $\Delta\alpha$ with NP (K_d = 54 nM and ΔH = -18.5 kcal/mol, Figure 11). This suggests that similar molecular interactions are involved in the complex formation of importin $\Delta\alpha$ with the binding motifs of the two histone types and of NP.

Importantly, ITC together with fluorescence anisotropy and centrifugation in sucrose gradients show that NP, histones and importin α can associate and form co-complexes, NP/H5/importin $\Delta\alpha$ and NP/H2A/H2B/importin $\Delta\alpha$ of discrete size, that would support a co-transport of histones and NP to the nucleus, mediated by the classical import pathway. Depending on the histone type, linker or core, and the amount of bound histones, different number of importin Δα molecules can be loaded on NP/histone complexes, in which Δα can bind both to NP-NLS as to histones-NLS-like binding sites, as was demonstrated using a NP mutant with impaired binding to Δα. The binding is an enthalpy driven process and it is characterized by nanomolar affinity [16].

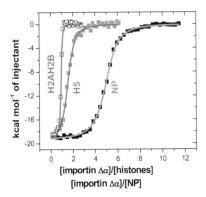

Figure 11. ITC isotherms for the binding interactions of the nucleosomal, H2A/H2B (violet) and linker, H5 (blue) histones, and NP (pink) with importin α ΔIBB (a truncated form lacking the autoinhibitory N-terminal domain)

We have also described the formation of quaternary NP/H5/importin α/β complexes by means of fluorescence and centrifugation, which makes conceivable that α/β heterodimer might "pull" NP/histones complexes into the nucleus, importin α binding either to NP, histones or both. Since importin β binds to both linker and core histones, NP/histones co-transport mediated by importin β, could also be expected. However, this hypothetical route seems unlikely since importin β competes with NP for histones, inducing the release of the latter from NP (unpublished data). Even though no detailed study has been performed, the comparable thermodynamic signature for H5/importin α and H5/importin α/β interaction supports the notion that H5 always binds through importin α in the presence of α/β heterodimer, which would explain the formation of quaternary complexes. Therefore the assembly of NP/histone/importin α/β complexes might have physiological meaning since it supports the existence of a putative and redundant histone import pathway in which positively charged histones would be protected against unspecific interactions by the histone chaperone nucleoplasmin.

7. Conclusion and future prospects

In summary, we have highlighted the importance of calorimetry in the study of nuclear chaperones. Detailed analysis demonstrated that the nuclear chaperone NP can associate with the two histone types and the transport machinery, and that co-complexes of NP, histones, and importins can assemble proving that ITC is suitable to study biological recognition in complex macromolecular assemblies. Notably, a link between NP phosphorylation state, its stability and the strength with which it assembles with histones is demonstrated.

One key feature of NP assembly with histones is the negative cooperative interactions, that render the protein operative in a wider concentration range and is an effective mechanism of regulation of the activity of macromolecular complexes. We have dissected the thermodynamic cooperativity of NP and its variants, core domain fragments and phosphorylation mimicking mutants, and presented strong evidence of the involvement of both NP domains in binding of histones, the NP tail domain being particularly essential in the assembly with H5 molecules. Only the nuclear localization signal NLS, however, is the recognition site in the multi-component NP/importin α/β complex. The significant differences in the enthalpic and entropic terms of the Gibbs free energy of NP association with histones and importins reflect different energetic strategy for NP chaperoning functions and its recognition for nuclear trafficking.

Both the experimental results and the methodological approach, ITC complemented with SAXS, allow a mechanistic understanding of nucleosome assembly/disassembly and its nuclear trafficking. The NP/histone complexes, which were modeled using five-fold symmetry, have a much more compact shape than the NP/importin α/β complex, reconstructed with multiple models, reflecting inherent flexibility.

Future work should focus towards description of the energetics of NPM export mechanism and the molecular recognition between NPM and nuclear export machinery (exportin), as well as with other proteins and peptides/small molecules. NPM is overexpressed in solid cancers (gastric, colon, ovarian and prostate), while genetic modifications of NPM1 gene by chromosomal translocation, mutation and deletion are involved in lymphomas and leukemias [102-105]. Mutations of NPM1 gene result in aberrant cytoplasmic localization of NPM in about 35% of acute myeloid leukemia (AML) patients [102]. The involvement of NPM in human cancer has received an increasing research interest during the last years, but the molecular mechanism of NPM implication in leukaemia and tumorigenesis is not understood yet. Studying the energetics of NPM binding with different (de)stabilizing ligands/drugs would help to regulate its interaction with cellular partners and thereby control its localization and function. This will entail, on one hand knowledge about the NPM nucleo-cytoplasmic shuttling and on the other is expected to provide a strategy for molecular therapeutics.

Author details

Stefka G. Taneva

Unidad de Biofísica (CSIC/UPV-EHU), Departamento de Bioquímica y Biología Molecular,
Universidad del País Vasco, Bilbao, Spain,
Institute of Biophysics and Biomedical Engineering, Bulgarian Academy of Sciences, Sofia, Bulgaria

Sonia Bañuelos and María A. Urbaneja
Unidad de Biofísica (CSIC/UPV-EHU), Departamento de Bioquímica y Biología Molecular,
Universidad del País Vasco, Bilbao, Spain

8. References

[1] Prado A, Ramos I, Frehlick LJ, Muga A, Ausió J (2004) Nucleoplasmin: a nuclear chaperone. Biochem. Cell Biol. 82, 437-445.

[2] Frehlick LJ, Eirín-López JM, Ausió J (2007) New insights into the nucleophosmin/nucleoplasmin family of nuclear chaperones. Bioessays 29, 49-59.

[3] Dutta S, Akey IV, Dingwall C, Hartman KL, Laue T, Nolte RT, Head JF, Akey CW (2001) The crystal structure of nucleoplasmin-core: implications for histone binding and nucleosome assembly. Mol. Cell 8, 841-853.

[4] Hierro A, Arizmendi JM, Bañuelos S, Prado A, Muga A (2002) Electrostatic interactions at the C-terminal domain of nucleoplasmin modulate its chromatin decondensation activity. Biochemistry 41, 6408-6413.

[5] Hierro A, Arizmendi JM, De las Rivas J, Urbaneja MA, Prado A, Muga A (2001) Structural and functional properties of Escherichia coli-derived nucleoplasmin. A comparative study of recombinant and natural proteins. Eur. J. Biochem. 268, 1739-1748.

[6] Sickmeier M, Hamilton JA, Le Gall T, Vacic V, Cortese MS, Tantos A, Szabo B, Tompa P, Chen J, Uversky VN, Obradovic Z, Dunker AK (2007) DisProt: the Database of Disordered Proteins. Nucleic Ac. Res. 35, D787 – D793.

[7] Cotten M, Sealy L, Chalkley R (1986) Massive phosphorylation distinguishes Xenopus laevis nucleoplasmin isolated from oocytes or unfertilized eggs. Biochemistry 25, 5063–5069.

[8] Bañuelos S, Omaetxebarria MJ, Ramos I, Larsen MR, Arregi I, Jensen OM, Arizmendi JM, Prado A, Muga A (2007) Phosphorylation of both nucleoplasmin domains is required for activation of its chromatin decondensation activity. J. Biol. Chem. 282, 21213-21221.

[9] Taneva SG, Muñoz I, Franco G, Falces J, Arregi I, Muga A, Montoya G, Urbaneja MA, Bañuelos S (2008) Activation of nucleoplasmin, an oligomeric histone chaperone, challenges its stability. Biochemistry 47, 13897-13906.

[10] Taneva SG, Bañuelos S, Arregi I, Falces J, Konarev P, Svergun D, Velázquez-Campoy A, Urbaneja MA (2009). A mechanism for histone chaperoning activity of nucleoplasmin: thermodynamic and structural models. J. Mol. Biol. 393, 448-463.

[11] Görlich D, Kutay U (1999) Transport between the cell nucleus and the cytoplasm. Annu. Rev. Cell Dev. Biol. 15, 607–660.

[12] Stewart M (2007) Molecular mechanism of the nuclear protein import cycle. Nat. Rev. Mol. Cell Biol. 8, 195–208.

[13] Fontes M R, Teh T, Kobe B (2000) Structural basis of recognition of monopartite and bipartite nuclear localization sequences by mammalian importin-α. J. Mol. Biol. 297, 1183–1194.

[14] Catimel B, Teh T, Fontes M R, Jennings IG, Jans DA, Howlett GJ, Nice EC, Kobe B (2001) Biophysical characterization of interactions involving importin-α during nuclear import. J. Biol. Chem. 276, 34189–34198.

[15] Falces J, Arregi I, Konarev P, Urbaneja MA, Taneva SG, Svergun D, Bañuelos S (2010) Recognition of nucleoplasmin by the nuclear transport receptor importin α/β: Insights into a complete transport complex. Biochemistry 49, 9756-9769.

[16] Arregi I, Falces J, Bañuelos S, Urbaneja MA, Taneva SG (2011) The nuclear transport machinery recognizes nucleoplasmin–histone complexes. Biochemistry 50, 7104-7110.

[17] Sánchez-Ruiz JM (1995) Differential scanning calorimetry of proteins. Subcell. Biochem. 24, 133–176.

[18] Jelesarov I, Bosshard HR (1999) Isothermal titration calorimetry and differential scanning calorimetry as complementary tools to investigate the energetics of biomolecular recognition. J. Mol. Recognit.12, 3-18.

[19] Cooper A (1999) Thermodynamics of protein folding and stability. In: Allen G, editor. Protein: A Comprehensive Treatise. JAI Press Inc.Volume 2, pp. 217-270.

[20] Schön A, Velázquez-Capmoy A (2005) Calorimetry In: Jiskoot W, Crommelin DJA, editors. Methods for structural analysis of protein pharmaceuticals. Arlington, VA, AAPS Press. pp. 573-589.

[21] Freire E, Mayorga OL, Straume M (1990) Isothermal titration calorimetry. Anal. Biochem. 179, 131-137.

[22] Laldbury JE, Chowdhry BZ (1996) Sensing the heat: the application of isothermal titration calorimetry to thermodynamic studies of biomolecular interactions. Chem. Biol. 3, 791-801.

[23] Pierce MM, Raman CS, Nall BT (1999) Isothermal titration calorimetry of protein-protein interactions. METHODS 19, 213-221.

[24] Leavitt S, Freire E (2001) Direct measurement of protein binding energetics by isothermal titration calorimetry. Curr. Opin. Struct. Biol. 11, 560-566.

[25] Cliff MJ, Ladbury J (2003) A survey of the year 2002 literature on applications of isothermal titration calorimetry. J. Mol. Recognit. 16, 383-391.

[26] Bjelic S, Jelesarov I (2008) A survey of the year 2007 literature on applications of isothermal titration calorimetry. J. Mol. Recognit. 21, 289-312.

[27] Falconer RJ, Collins BM (2009) Survey of the year 2009: applications of isothermal titration calorimetry. J. Mol. Recognit. 24, 1-16.

[28] Ghai R, Falconer RJ, Collins BM (2011) Applications of isothermal titration calorimetry in pure and applied research - survey of the literature from 2010. J. Mol. Recognit. 25, 32-52.

[29] Freire E (1995) Thermal denaturation methods – study of protein folding. Methods in Enzymol. 259, 144-168.

[30] Brown A (2009) Analysis of cooperativity by isothermal titration calorimetry. Int. J. Mol. Sci. 10, 3457-3477.

[31] Freire E, Schön A, Velázquez-Campoy A (2009) Isothermal titration calorimetry: general formalism using binding polynomials. Methods Enzymol. 455, 127-155.

[32] Velázquez-Campoy A, Goñi G, Peregrina JR, Medina M (2006) Exact analysis of heterotropic interactions in proteins: characterization of cooperative ligand binding by isothermal titration calorimetry. Biophys. J. 91, 1887-1904.

[33] Martinez-Julvez M, Abian O, Vega S, Medina M, Velázquez-Campoy A (2012) Studying the allosteric energy cycle by isothermal titration calorimetry. Methods in Molecular Biology 796, 53-70.

[34] Shiou-Ru T, Charalampos GK (2009) Dynamic activation of an allosteric regulatory protein. Nature 462, 368-374.

[35] Moro F, Taneva SG, Velázquez-Campoy A, Muga A (2007) GrpE N-terminal domain contributes to the interaction with DnaK and modulates the dynamics of the chaperone substrate binding domain. J. Mol. Biol. 374, 1054-1064.

[36] Taneva S, Moro F, Velázquez-Campoy A, Muga A (2010) Energetics of nucleotide-induced DnaK conformational states. Biochemistry 49, 1338–1345.

[37] Seldeen KL, Deegan BJ, Bhat V, Mikles DC, McDonald CB, Farooq A (2011) Energetic coupling along an allosteric communication channel drives the binding of Jun-Fos heterodimeric transcription factor to DNA. FEBS J. 278, 2090-2104.

[38] Sturtevant JM (1977) Heat capacity and entropy changes in processes involving proteins. Proc. Natl. Acad. Sci. U.S.A. 74, 2236-2240.

[39] Spolar RS, Record MT (1992) Coupling of local folding to site-specific binding of proteins to DNA. Science 263, 777-784.

[40] Makhatadze GI, Privalov PL (1995) Energetics of protein structure. Advan. Protein Chem. 47, 307-425.

[41] Murphy KP, Freire E (1992) Thermodynamics of structural stability and cooperative folding behavior in proteins. Adv. Protein Chem. 43, 313-336.

[42] Murphy KP, Xie D, Garcia KC, Amzel LM, Freire E (1993) Structural energetics of peptide recognition: angiotensin II/antibody binding. Proteins: Struct. Funct. Genet. 15, 113-120.

[43] Ladbury JE, Wright JG, Sturtevant JM, Sigler PB (1994) A thermodynamic study of the trp repressor-operator interaction. J. Mol. Biol. 238, 669-681.

[44] Pearce KHJr, Ultsch MH, Kelley RF, de Vos AM, Wells JA (1996) Structural and mutational analysis of affinity-inertcontact residues at the growth hormone-receptor interface. Biochemistry 35, 10300-10307.

[45] Holdgate GA, Tunnicliffe A, Ward WH, Weston SA, Rosenbrock G, Barth PT, Taylor IW, Pauptit RA, Timms D (1997) The entropic penalty of ordered water accounts for

weaker binding of the antibiotic novobiocin to a resistant mutant of DNA gyrase: a thermodynamic and crystallographic study. Biochemistry 36, 9663-9673.

[46] Baker BM, Murphy KP (1996) Evaluation of linked protonation effects in protein binding reactions using isothermal titration calorimetry. Biophys. J. 71, 2049-2055.

[47] Baldwin RL (1986) Temperature dependence of the hydrophobic interaction in protein folding. Proc. Natl. Acad. Sci. U.S.A. 83, 8069–8072.

[48] Murphy KP, Xie D, Thompson KS, Amzel LM, Freire E (1990) Entropy in biological binding processes: estimation of translational entropy loss. Science 247, 559–561.

[49] Luque I, Freire E (1998) Structure-based prediction of binding affinities and molecular design of peptide ligands. Methods Enzymol. 295, 100–127.

[50] Gómez J, Freire E (1995) Thermodynamic mapping of the inhibitor site of the aspartic protease endothiapepsin. J. Mol. Biol. 252, 337-350.

[51] Fukada H, Takahashi K (1998) Enthalpy and heat capacity changes for the proton dissociation of various buffer components in 0.1 M potassium chloride. Proteins: Struct. Func. Genet. 33, 159-166.

[52] Petrucci S (1972) Ionic Interactions 1, 117–177.

[53] Kholodenko V, Freire E (1999) A simple method to measure the absolute heat capacity of proteins. Anal. Biochem. 270, 336–338.

[54] Sánchez-Ruiz JM, López-Lacomba JL, Cortijo M, Mateo PL (1988) Differential scanning calorimetry of the irreversible thermal denaturation of thermolysin. Biochemistry 27, 1648-1672.

[55] Sánchez-Ruiz JM (1992) Theoretical analysis of Lumry-Eyring models in differential scanning calorimetry. Biophys. J. 61, 921-935.

[56] Kurganov BI, Lyubarev AE, Sánchez-Ruiz JM, Shnyrov VL (1997) Analysis of differential scanning calorimetry data for proteins. Criteria of validity of one-step mechanism of irreversible protein denaturation. Biophys. Chem. 69, 125–135.

[57] Freire E, Murphy KP, Sánchez-Ruiz JM, Galisteo ML, Privalov PL (1992) The molecular basis of cooperativity in protein folding. Thermodynamic dissection of interdomain interactions in phosphoglycerate kinase. Biochemistry 31, 250-256.

[58] Galisteo ML, Sánchez-Ruiz JM (1993) Kinetic study into the irreversible thermal denaturing of bacteriorhodopsin. Eur. Biophys. J. 22, 25-30.

[59] Landin JS, Katragadda M, Albert A (2001) Thermal destabilization of rhodopsin and opsin by proteolytic cleavage in bovine rod outer segment disk membranes. Biochemistry 40, 11176-11183.

[60] Milardi D, La Rosa C, Grasso D, Guzzi RC, Sportelli L, Carlo F (1998) Thermodynamic and kinetics of the thermal unfolding of plastocyanin. Eur. Biophys. J. 27, 273-282.

[61] Krumova SB, Todinova SJ, Busheva MC, Taneva SG (2005) Kinetic nature of the thermal destabilization of LHCII Macroaggregates. J. Photochem. Photobiol. B 78, 165-170.

[62] Guzzi R, La Rosa C, Grasso D, Milardi D, Sportelli L (1996) Experimental model for the thermal denaturation of azurin: a kinetic study. Biophys. Chem. 60, 29–38.

[63] Meijberg W, Schuurman-Wolters GK, Boer H, Scheck RM, Robillard GT (1998) The thermal stability and domain interactions of the mannitol permease of *Escherichia coli*. A differential scanning calorimetry study. J. Biol. Chem. 273, 20785-20794.

[64] Lubarev AE, Kurganov BI (2001) Study of irreversible thermal denaturation of proteins by differential scanning calorimetry. Recent Res. Devel. Biophys. Chem. 2, 141-165.

[65] Davoodi J, Wakarchuk WW, Surewicz WK, Carey PR (1998) Scan-rate dependence in protein calorimetry: The reversible transitions of *Bacillus circulans* xylanase and a disulfide-bridge mutant. Protein Sci 7, 1538-1544.

[66] Brandts JF, Lin L-N (1990) Study of strong to ultratight protein interactions using differential scanning calorimetry. Biochemistry 29, 6927-6940.

[67] Garbett NC, Miller JJ, Jenson AB, Miller DM, Chaires JB (2007) Interrogation of the plasma proteome with differential scanning calorimetry. Clin. Chem. 53, 2012-2014.

[68] Garbett NC, Miller JJ, Jenson AB, Chaires JB (2008) Calorimetry outside the box: A new window into the plasma proteome. Biophys. J. 94, 1377–1383.

[69] Garbett NC, Mekmaysy C, Helm CV, Jenson AB, Chaires JB (2009) Differential scanning calorimetry of blood plasma for clinical diagnosis and monitoring. Exp. Mol. Pathol. 86, 186-191.

[70] Todinova S, Krumova S, Gartcheva L, Robeerst C, Taneva SG (2011) Microcalorimetry of blood serum proteome – a modified interaction network in the multiple myeloma case. Anal. Chem. 83, 7992-7998.

[71] Ruben AJ, Kiso Y, Freire E (2006) Overcoming roadblocks in lead optimization: A thermodynamic perspective. Chem. Biol. Drug Des. 67, 2–4.

[72] Cai L, Cao A, Lai L (2001) An isothermal titration calorimetric method to determine the kinetic parameters of enzyme catalytic reaction by employing the product inhibition as probe. Anal. Biochem. 299, 19–23.

[73] Chaires JB (2006) A thermodynamic signature for drug–DNA binding mode. Arch. Biochem. Biophys. 453, 26–31.

[74] Burnouf D, Ennifar E, Guedich S, Puffer B, Hoffmann G, Bec G, Disdier F, Baltzinger M, Dumas P (2012) kinITC: A new method for obtaining joint thermodynamic and kinetic data by isothermal titration calorimetry. J. Am. Chem. Soc. 134, 559-565.

[75] Reymond C, Bisaillon M, Perreault J-P (2009) Monitoring of an RNA multistep folding pathway by isothermal titration calorimetry. Biophys. J. 96, 132-140.

[76] Kardos J, Yammamoto K, Hasegawa K, Naiki H, Goto Y (2004) Direct measurement of thermodynamic parameters of amyloid formation by isothermal titration calorimetry. J. Biol. Chem. 279, 55308-55314.

[77] Zhou X, Manjunatha K, Sivaraman J (2011) Application of isothermal titration calorimetry and column chromatography for identification of biomolecular targets. Nature protocols 6, 158-165.

[78] Franco G, Bañuelos S, Falces J, Muga A, Urbaneja MA (2008) Thermodynamic characterization of nucleoplasmin unfolding: interplay between function and stability. Biochemistry 47, 7954-7962.

[79] Vancurova S, Paine TM, Lu W, Paine PL (1995) Nucleoplasmin associates with and is phosphorylated by casein kinase II. J. Cell Sci. 108,779–787.

[80] Bañuelos S, Hierro A, Arizmendi JM, Montoya G, Prado A, Muga A (2003) Activation mechanism of the nuclear chaperone nucleoplasmin: role of the core domain. J. Mol. Biol. 334, 585-593.

[81] Tholey A, Lindermann A, Kinzel V, Reed J (1999) Direct effects of phosphorylation on the preferred backbone conformation of peptides: a nuclear magnetic resonance study. Biophys. J. 76, 76–87.

[82] Philpott A, Leno GH (1992) Nucleoplasmin remodels sperm chromatin in *Xenopus* egg extracts. Cell 69, 759-767.

[83] Koshland DEJr (1996) The structural basis of negative cooperativity: receptors and enzymes. Curr. Opin. Struct. Biol. 6, 757-761.

[84] Schön A, Freire E (1989) Thermodynamics of intersubunit interactions in cholera toxin upon binding to the oligosaccharide portion of its cell surface receptor, ganglioside GM1. Biochemistry, 28, 5019–5024.

[85] Conti E, Uy M, Leighton L, Blobel G, Kuriyan J (1998) Crystallographic analysis of the recognition of a nuclear localization signal by the nuclear import factor karyopherin alpha. Cell 94, 193–204.

[86] Conti E, Kuriyan J (200) Crystallographic analysis of the specific yet versatile recognition of distinct nuclear localization signals by karyopherin alpha. Structure 8, 329–338.

[87] Fanara P, Hodel MR, Corbett AH, Hodel AE (2000) Quantitative analysis of nuclear localization signal (NLS)-importin alpha interaction through fluorescence depolarization. Evidence for auto-inhibitory regulation of NLS binding. J. Biol. Chem. 275, 21218–21223.

[88] Fontes MR, Teh T, Jans D, Brinkworth RI, Kobe B (2003) Structural basis for the specificity of bipartite nuclear localization sequence binding by importin-α. J. Biol. Chem. 278, 27981–27987.

[89] Yang SN, Takeda AA, Fontes MR, Harris JM, Jans DA, Kobe B (2010) Probing the specificity of binding to the major nuclear localization sequence-binding site of importin-α using oriented peptide library screening. J. Biol. Chem. 285, 19935–19946.

[90] Kobe B (1999) Autoinhibition by an internal nuclear localization signal revealed by the crystal structure of mammalian importin. R. Nat. Struct. Biol. 6, 388–397.

[91] Harreman MT, Cohen PE, Hodel MR, Truscott GJ, Corbett AH, Hodel AE (2003) Characterization of the auto-inhibitory sequence within the N-terminal domain of importin alpha. J. Biol. Chem. 278, 21361–21369.

[92] Tarendeau F, Boudet J, Guilligay D, Mas PJ, Bougault CM, Boulo S, Baudin F, Ruigrok RW, Daigle N, Ellenberg J, Cusack S, Simorre JP, Hart DJ (2007) Structure and nuclear import function of the C-terminal domain of influenza virus polymerase PB2 subunit. Nat. Struct. Mol. Biol. 14, 229–233.

[93] Dias SM, Wilson KF, Rojas KS, Ambrosio AL, Cerione RA (2009) The molecular basis for the regulation of the cap-binding complex by the importins. Nat. Struct. Mol. Biol. 16, 930–937.

[94] Dingwall C, Sharnick SV, Laskey RA (1982) A polypeptide domain that specifies migration of nucleoplasmin into the nucleus. Cell 30, 449-458.

[95] Jans DA (1995) The regulation of protein transport to the nucleus by phosphorylation. Biochem. J. 311, 705-716.

[96] Hood JK, Silver PA (1999). In or out? Regulating nuclear transport. Curr. Opin. Cell Biol. 11, 241-247.

[97] Fontes MR, Teh T, Toth G, John A, Pavo I, Jans DA, Kobe B (2003) Role of flanking sequences and phosphorylation in the recognition of the simian-virus-40 large T-antigen nuclear localization sequences by importin-α. Biochem. J. 375, 339-349.

[98] Cingolani G, Petosa C, Weis K, Muller CW (1999) Structure of importin-β bound to the IBB domain of importin-alpha. Nature 399, 221–229.

[99] Jäkel S, Albig W, Kutay U, Bischoff FR, Schwamborn K, Doenecke D, Görlich D (1999) The importin beta/importin 7 heterodimer is a functional nuclear import receptor for histone H1. EMBO J. 18, 2411-2423.

[100] Jäkel S, Mingot JM, Schwarzmaier P, Hartmann E, Görlich D (2002) Importins fulfil a dual function as nuclear import receptors and cytoplasmic chaperones for exposed basic domains. EMBO J. 21, 377-386.

[101] Johnson-Saliba M, Siddon NA, Clarkson MJ, Tremethick DJ, Jans DA (2000) Distinct importin recognition properties of histones and chromatin assembly factors. FEBS Lett. 467, 169-174.

[102] Falini B, Mecucci C, Tiacci E, Alcalay M, Rosati R, Pasqualucci L, La Starza R, Diverio D, Colombo E, Santucci A, Bigerna B, Pacini R, Pucciarini A, Liso A, Vignetti M, Fazi P, Meani N, Pettirossi V, Saglio G, Mandelli F, Lo-Coco F, Pelicci PG, Martelli MF (2005) Cytoplasmic nucleophosmin in acute myelogenous leukemia with a normal karyotype.Engl. J. Med. 352, 254-266.

[103] Falini B, Bigerna B, Pucciarini A, Tiacci E, Mecucci C, Morris SW, Bolli N, Rosati R, Hanissian S, Ma Z, Sun Y, Colombo E, Arber DA, Pacini R, La Starza R, Verducci Galletti B, Liso A, Martelli MP, Diverio D, Pelicci PG, Lo Coco F, Martelli MF (2006) Aberrant subcellular expression of nucleophosmin and NPM-MLF1 fusion protein in acute myeloid leukaemia carrying t(3;5): a comparison with NPMc+ AML. Leukemia 20, 368-371.

[104] Falini B, Bolli N, Liso A, Martelli MP, Mannucci R, Pileri S, Nicoletti I (2009) Altered nucleophosmin transport in acute myeloid leukaemia with mutated NPM1: molecular basis and clinical implications. Leukemia 23,11731-1743.

[105] Grisendi S, Mecucci C, Falini B, Pandolfi PP (2006) Nucleophosmin and cancer. Nature Reviews 6, 493-505.

Application of MicroCalorimetry to Study Protein Stability and Folding Reversibility

Calorimetric Study of Inulin as Cryo- and Lyoprotector of Bovine Plasma Proteins

Laura T. Rodriguez Furlán, Javier Lecot, Antonio Pérez Padilla, Mercedes E. Campderrós and Noemi E. Zaritzky

Additional information is available at the end of the chapter

1. Introduction

Inulin is a generic term applied to heterogeneous blends of fructo-oligosaccharides [1] which are reserve carbohydrate sources present in many plant foods such as bananas, onions, garlic, leeks, artichokes and chicory, which represents the main commercial source. This polysaccharide has a wide range of both, nutritional and technological applications. Nutritionally, inulin is regarded as a soluble fiber which promotes the growth of intestinal bacteria, acting as a prebiotic. Also, is a non-digestible carbohydrate with minimal impact on blood sugar and unlike fructose, it is not insulemic and does not raise triglycerides being generally considered suitable for diabetics and potentially helpful in managing blood sugar-related illnesses [2-4]. Among the technological benefits, inulin is used as fat and sugar replacement, low caloric bulking agent, texturing and water-binding agent [5,6]. One general property of the saccharides is the stabilization of proteins by their incorporation into carbohydrate solutions before freeze-drying being this a known preservation procedure [7-10]. The previous incorporation of saccharide promotes the formation of amorphous, glassy systems, inhibits crystallization and influences the kinetics of deteriorative reactions upon storage by which its structured integrity is maintained [8,9,11,12]. To act successfully as a protectant, the saccharides should have a high glass transition temperature (T_g), a poor hygroscopicity, a low crystallization rate, containing no reducing groups. When freeze-drying is envisaged as a method of drying, a relatively high T'_g of the freeze concentrated fraction is preferable. Previous studies demonstrated that inulin meets these requirements being excellent protector of therapeutical proteins and viruses over the drying and storage processes [13,14].

The protein preserved by freeze-drying simplifies aseptic handling and enhances stability of protein products, with limited shelf lives in solution, by obtaining a dry powder without excessive heating. However, during the freeze-drying process the protein may lose its activity and must be protected from conformational changes or denaturation [11,15]. The stabilization of proteins conferred by saccharides during freeze- drying has been explained by several mechanisms. First, replacing the hydrogen bonding between water and protein stabilizes the protein during drying processes, and second, the formation of a glass matrix where the protein is encapsulated avoiding its unfolding and thus preserving its conformation during freeze-drying [8,12,16-18]. Therefore, through the correct selection of the saccharide it is possible to improve the stability of proteins through their encapsulation in a glassy matrix, where molecular mobility is quite limited so that the rates of diffusion-controlled reactions, like protein unfolding or protein aggregation, are reduced [16,19,20].

Information about the energy of a protein can be obtained by means of thermal denaturation studies, allowing the characterization of their behavior during freeze-drying cycle. Differential scanning calorimetry (DSC) is one of the most useful methods for assessing protein thermal behavior and to obtain thermodynamic parameters of folding-unfolding transitions [21].

During the freeze-drying of a protein solution with or without saccharides to protect the structure, the primary drying is the most time consuming stage of the process. It should be carry out at the maximum allowable temperature usually associated to the glass transition temperature of the maximally freeze concentrate solution (T'_g). Below this temperature a glassy state that behaves as an amorphous solid is obtained. If the temperature of the frozen system rises above the T'_g, the material becomes less viscous and freeze-drying may cause the loss of the porous structure and the product collapse [20,22,23]. In the freeze-dried sample, water is removed and the solute concentration in the matrix increases, obtaining a material with an amorphous structure that exhibits a glass-rubber transition at a specific temperature which is named as the glass transition temperature (T_g) [24-28]. It is noteworthy that amorphous materials are stable in the glassy state below T_g, when the temperature is higher the viscosity decreases and thus the rate of chemical reactions increases and crystallization events occur, increasing the rate of deterioration during storage [22,25,27-29]. Both transitions T'_g and T_g are important parameters in the development of the freeze-drying cycle because not only ensures the stability and quality of the product, but also allow to improve the efficiency of the manufacturing process [20,22,28,30].

A diagram of phases for the water-saccharide system is shown in Figure 1. The curve of the freezing temperature separates the zones corresponding to the liquid and the solid (ice) solution phases. In fact, this procedure is aimed at obtaining a glassy system at room temperature as indicated in D. To get to this state, the freeze-dried process indicated by the curve A-B-C-D-E is carried out. The curve for the glass transition temperature (T_g) is reached when the solution is overcooled (B-C) until the T'_g in point C, where the concentration of the vitrification agent (saccharide) is given by C'_g. Then the water is eliminated and the solute concentration increases (C-D-E), obtaining a solid with an amorphous structure that exhibits a glass transition temperature (T_g) [22,28].

Therefore, the determination of the freeze-drying cycle is important because of physical changes that occur in the solution during the process, its study can be applied to improve processability, quality, and stability of the product during storage [29].

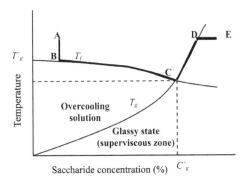

Figure 1. Phases diagram of the water–saccharide system. The curve A-B-C-D-E indicated the freeze-dried process. (T_f = freezing temperature; T_g = glass transition temperature; T'_g = glass transition temperature of the overcooling solution; C'_g = saccharide concentration)

Although many authors reported the use of saccharides as cryoprotectants of proteins and inulin as a good protector agent of some compounds, the present study is an attempt to evaluate inulin as cryoprotector of food proteins such as bovine plasma proteins, taking profit of the nutritional and technological benefit of the polysaccharide. Also there is a limited amount of data on glass transition temperatures for multicomponent mixtures and on the comparison of experimental and predicted values for such mixtures [28]. Then, the purposes of this study were *i)* to investigate the transition temperatures and the thermal denaturation of bovine plasma proteins stabilized with inulin in a glassy matrix in comparison to the effect of a monosaccharide (glucose) and a disaccharide (sucrose) at different concentrations using DSC, *ii)* to compare the quality, performance and storage conditions of these products.

The glass transition temperatures of the maximally concentrated frozen solutions (T'_g) were analyzed and compared to the experimental results by applying the predictive equations of Miller/Fox and Gordon/Taylor extended for multi-component systems. The glass transition (T_g) of the freeze dried multi-component mixtures, the onset crystallization temperature (T_c) of the solute at temperatures above T_g, in the freeze dried samples were determined. Furthermore, the kinetics of the denaturation and the thermal denaturation (T_d) of the freeze-dried samples, at different DSC scan rates, protein concentrations and pH, were analyzed and the thermodynamic compatibility of the different matrix components were determined. The enthalpy of change involved in the denaturation reactions of proteins (ΔH) was also determined. A kinetic model that describes bovine plasma proteins denaturation was proposed.

2. Materials and methods

2.1. Raw materials

The inulin used as cryoprotectant is mainly constituted by linear chains of fructose, with a glucose terminal unit, and has a molecular weight of 2400 g/mol. The commercial product was provided by Orafti Chile S.A. and was obtained from chicory. The other saccharides employed to compare their performance were: *i)* a monosaccharide, glucose (Parafarn, Argentine), with a purity of 99.99% and *ii)* a disaccharide, commercial sucrose (Ledesma S.A., Argentine).

The protein used in the study was spray dried bovine plasma (Yerubá S.A. Argentine). The molecular weights of the proteins were in the range of 15.000 to 80.000 Da. The composition was 76±5% proteins, <0.1% fat, 10% ash, 4% water, 1% low molecular weight compounds.

2.2. Preparation of Protein/carbohydrate samples: Concentration of bovine plasma proteins through ultrafiltration and freeze-drying treatments

The protein concentrate was obtained by means of a membrane process, which allowed protein concentration, eliminating insoluble macroscopic components, reducing the saline content [18]. The steps of the process were: i) the bovine plasma was dissolved in de-ionized water to a concentration of 3% w/v using a mixer at a low speed to avoid the formation of vortex and to minimize the appearance of foam; ii) the solution was passed through a porous support (Viledon FO 2431D, Germany) to remove macroscopic aggregates and reduce the saline content; iii) the feed solution (3 L) was thermostatized in a water bath and impelled with a centrifugal pump, first through a frontal flow stainless steel filter, with a pore size of 60 μm (Gora, Argentine) (this procedure of microfiltration (MF) reduces the amount of bacteria and spores and acts as cold pasteurization, moreover this stage protects the ultrafiltration (UF) membrane from fouling); and finally, iv) the UF was performed using Pellicon cassette module (Millipore, Bedford, MA, USA), containing modified polyethersulfone membranes with a molecular weight cut-off (MWCO) of 10 kDa, with a membrane area of 0.5 m². The concentration of proteins by UF was carried out by continuously removing the permeate stream until the desired concentration of 4% (w/v), was achieved. The experimental runs were performed at a transmembrane pressure (ΔP) of 1.5 bar, flow rate of (2.9 ± 0.05) L/min and a temperature of 10 °C. Additionally a discontinuous diafiltration (DD) process was applied to removal salts and other contaminant of low molecular weight. For this operation the starting solution was the UF concentrate, which was diluted to the initial volume (3 L) with de-ionized water in a single state and ultrafiltered to the desired concentration range.

The UF membrane undergoes a fouling process during protein permeation so a cleaning protocol may be applied. It was performed by applying a "Cleaning in Place" (CIP) procedure according to the manufacturer's instructions. At the end of each run, a cycle of water/ alkali (NaOH, pH=12.5 ± 0.5)/ water wash was applied to the membrane at (40 ± 2) °C and at a transmembrane pressure of 1 bar. Furthermore, a cleaning step using NaClO (commercial grade) 300 ppm was carried out at the same temperature and pressure to ensure sanitation and cleaning. Measurements of normalized water permeability were performed in order to verify recovery of flow through the membrane which ensures the recuperation of membrane permeability.

The bovine plasma protein (BPP) concentrate obtained by UF (concentration: 4 % w/v) was fractioned: A fraction as witness sample was reserved and the protective agent (glucose, sucrose, inulin) was added to the rest, in concentrations of 5%, 10% and 15% (w/v). A part of these solutions was reserved for DSC analysis to determine T'_g and the others were placed on stainless steel trays, frozen in a freezer at -40 °C and freeze-dried using a lyophilizer (Rificor S.A., Argentine) at 1 bar for 48 h. The samples temperature was controlled by a temperature sensor. The denatured protein content was determined before and after the freeze-drying.

2.3. Differential Scanning Calorimetry (DSC) measurements

Determination of T'_g in the protein solutions

The solutions containing plasma proteins–saccharides mixture were analyzed to determine T'_g at different pH values and saccharide concentrations by DSC with a Q100DTA Instrument (USA). The pH was adjusted using 0.1 N of NaOH and HCl. Protein concentrate solutions (average composition: saccharide 5% p/v - protein 4% p/v; saccharide 10% p/v - protein 4% p/v; saccharide 15% p/v - protein 4% p/v), (10 ± 2 mg) were weighed into aluminum DSC pans, hermetically sealed, and then loaded onto the DSC instrument at room temperature, using an empty pan as a reference. Samples Solutions were: (a) equilibrated at 20 °C and held for 1 min; (b) cooled at 2 °C/min until -80 °C for glucose, -60 °C for sucrose and -40 °C for inulin and held for 30 min; (c) warmed up to the annealing temperature (-50, -40 and -20 °C, for glucose, sucrose and inulin, respectively) by employing an annealing time of 30 min at heating rate of 2 °C/min [31]; (e) recooled at the same temperature of step (b) and held for 30 min; (f) warmed up to 0 °C at heating rate of 2 °C/min. The effectiveness of the procedure was verified corroborating the absence of ice devitrification in thermograms, that is to say the nonexistence of an exothermic peak previous to the ice melting.

Determination of T_g, T_c and T_d of proteins in the freeze–dried solids

Heat induced conformational changes on freeze-dried bovine plasma protein concentrate (BPP concentrate) in the amorphous carbohydrate matrix. The freeze-dried solids were analyzed to determine T_g, T_c and T_d at different pH values and saccharide concentrations by DSC with a Q100DTA Instrument (USA). The pH was adjusted using 0.1 N of NaOH and HCl. Protein concentrates (average composition: freeze-dried with saccharide 5% (p/v) = saccharide 35% p/p - protein 55% p/p; freeze-dried with saccharide 10% (p/v) = saccharide 64% p/p - protein 28% p/p; freeze-dried with saccharide 15% (p/v) = saccharide 79 % p/p - protein 14% p/p), (12.5 ± 2.5 mg) were weighed into aluminum DSC pans, hermetically sealed, and then loaded onto the DSC instrument at room temperature, using an empty pan as a reference.

Freeze–dried solids were equilibrated at 0 °C, held for 1 min and then warmed up to 200 °C at heating rate of 2 °C/min. To check the irreversibility of the reaction of heat-induced conformational changes, the samples after the end of the first heating stage described before, were re-scanned. For this, the protein-saccharide samples were cooled to 20 °C and stabilized during 5 min, and then warmed up to 200°C. Samples of freeze dried bovine plasma protein concentrate (BPP concentrates) in the amorphous carbohydrate matrix at pH

8, 6 and 4, at different heating rates of 2 and 5 °C/min in the temperature range 20–200 °C were analyzed. The pH was adjusted using 0.1 N of NaOH and HCl. Measurements were carried out on three separate samples (replicates). The following parameters were calculated at least in triplicate: T_d, at maximum heat flow, and ΔH, the enthalpy change involved in the overall heat-induced reactions within the protein molecules, that was determined by integrating the area beneath the enthalpy peak and above a straight baseline drawn in between the beginning and the end of the transition temperature range [32-34]; the T'_g and T_g were determined from the midpoint of the transition of the baseline shift on the amorphous sample.

In the freeze dried samples, at temperatures above T_g, the onset crystallization temperature (T_c) of the added solute was determined from the intersection of the baseline and the tangent of the exothermic peak. The enthalpy change involved in the overall heat-induced reactions within the protein molecules, ΔH_c, was determined by integrating the area beneath the exothermic peak and above a straight baseline drawn between the beginning and end of the transition temperature range [22,32,33].

2.4. Determination of native protein content

The native protein content is a measure of protein functionality preservation. It was determined after isoelectric precipitation of denatured/aggregated protein [18,35]. Dispersions of protein concentrate at 1% (w/v) were adjusted to pH value inferior of the pI of plasma proteins (~ 4.8) using 0.1 N of NaOH and HCl. An aliquot of the solution was centrifuged in a refrigerated ultracentrifuge (Beckman J2-HS) at 20,000 rpm 30 min at 5 °C. Protein concentration in the supernatants was diluted in a dissociating buffer (EDTA 50 mM, urea 8 M, pH= 10) and determined by molecular absorptiometry at 280 nm. The results were reported as percentage of the total protein concentration [36]. The percentage of native protein content of suspensions at pH 4.8 was obtained as the ratio between soluble protein (SP) and total protein (TP) contents after aggregation of denatured protein (Eq. 1).

$$NP\% = \left(\frac{SP}{TP}\right) \times 100 \qquad (1)$$

2.5. Scanning electron microscopy

The microstructure of freeze-dried plasma concentrates with and without saccharides was analyzed by scanning electron microscopy (SEM) using an LEO1450VP equipment (Zeiss, Germany). Powder samples were mounted on double-sided carbon adhesive tape on aluminum stubs and gold-coated and processed in a standard sputter. The micrographs were obtained in high vacuum at 10 KeV.

2.6. Statistical analysis

The experimental data were statistically analyzed by the Tukey-Kramer multiple comparison test, in the cases where 2 or more comparisons were considered, assuming that a $P<0.05$ was statistically significant [37]. Statistical GraphPad InStat software (1998) was used.

3. Theoretical considerations

3.1. Equations for T'_g prediction

The Miller/Fox equation can be used for the determination of T'_g dependence with the composition in a multi-component system, assuming constant density of the solutions, independent of temperature [28,38,39]. For a ternary mixture (protein-saccharide-water), it can be written as:

$$\frac{1}{T_g} = \frac{m_1}{m_t T_{g1}\left(\rho_1 / \rho_t\right)} + \frac{m_2}{m_t T_{g2}\left(\rho_2 / \rho_t\right)} + \frac{m_3}{m_t T_{g3}\left(\rho_3 / \rho_t\right)} \tag{2}$$

where T_g, glass transition temperature; m, mass; ρ, density; the subscripts t, 1, 2, 3 mean: total and each pure component, respectively.

The Gordon and Taylor equation [40] predicts the plasticizing effect of water on the T_g for a multicomponent system. The equation has been used among others, for systems treated as binary mixtures, determining experimentally the glass transition of the respective solid [41,42]. Instead we proposed a system considering each individual component: bovine protein concentrate, saccharide and water, with each corresponding property [43]:

$$T_g = \frac{w_1 T_{g1} + k w_2 T_{g2} + k^2 w_3 T_{g3}}{w_1 + k w_2 + k^2 w_3} \tag{3}$$

where w_1, w_2, w_3, are the weight fraction of each component defined as (m_i / m_t), and k is an empirical constant proportional to the plasticizing effect of water. This parameter was calculated to fit experimental data from a nonlinear optimization procedure (Gauss Newton procedure) using the software Excel 2003 (Microsoft).

Eqs. (2) and (3) were used for the determination of T'_g of the frozen solutions.

3.2. Theory of protein unfolding

Unfolding of protein is suggested to involve at least two steps according to Lumry and Eyring model (1954). The first step is a reversible unfolding of the native protein (N). This is followed by an irreversible change of the denatured protein (D) into a final irreversible state (I) [44,45].

$$N \underset{k_{-1}}{\overset{k_1}{\leftrightarrow}} D \overset{k_2}{\rightarrow} I \tag{4}$$

A special case was when $k_2 \gg k_{-1}$, where most of the D molecules will be converted to I as an alternative to refolding back to the native state. In this case, the denaturation process can be regarded as one-step process following first-order kinetics [44-46], (Eq.5).

$$N \xrightarrow{k} I \qquad\qquad\qquad (5)$$

where the first-order rate constant k can be identified with k_1 of Eq. (4). The total absorbed heat now equals the enthalpy change from N to I; it was generally assumed that the enthalpy change from D to I was negligible compared to that from N to D [44].

Experimentally, the irreversibility of unfolding was verified in a rescan. For an irreversible process, in the DSC rescanned thermograms no transition could be observed.

4. Results and discussion

4.1. Effect of saccharides on glass transition of the freeze concentrated matrix

As was previously mentioned, to avoid collapse of the products during the freeze-dried process, a temperature below the glass transition temperature of the frozen concentrated solutions, must be attained. Inulin as protein protective agent was comparatively studied, employing mono and disaccharides. The thermograms of Figure 2 show the transition temperatures of the frozen solutions of bovine plasma with inulin compared to the other saccharides, obtained in a single scan.

The result indicated that at each saccharide concentration, T'_g was higher for inulin (Table 1), suggesting that it has a greater cryostabilizing effect on bovine plasma proteins than the other saccharides, improving product stability. It was also observed that T'_g increased with the molecular weight of the cryoprotectant that is: inulin > sucrose > glucose. The same tendency was reported previously by means of the evaluation of protein shelf life time [18]. By the other hand, it was reported that inulin exhibit better stabilizing properties than sucrose and trehalose in the prevention of the nonPEGlated lipoplexes aggregation [14]. Many studies concluded that transition temperatures increased with the saccharides molecular weight [22,23,27]. For example T'_g of freeze–dried surimi depended strongly on the type and content of sugar and at each sugar level the T'_g was trehalose > sucrose > glucose > sorbitol [47].

Thermograms of bovine plasma solutions revealed the existence of two glass transitions (T'_{g1} and T'_{g2}) for glucose and sucrose as protective agents, evidenced as deviations in the base line (indicated by arrows in Figure 1). Similar results were found by Telis and Sobral [48] who worked with freeze–dried tomato. This may be because the presence of phases formed by different proportions of saccharide, water, and proteins present in the frozen solution [47-49]. Also it was observed that when the saccharide concentration increased, T'_{g1} and T'_{g2} increased and decreased respectively (Table 1). However a constant average value was maintained between both T'_g values for each sugar, being -51.2 ± 0.8 and -41.1 ± 0.1 for glucose and sucrose, respectively. Similar results were found in [47] on freeze–dried surimi product with trehalose. For inulin only one T'_g was found, which increased with the increase of saccharide concentration. From these results and considering that the water acts as plasticizer, i.e. decreases drastically T'_g of food polymers [26], it can be concluded that the conditions of the freeze-drying process, are linked directly to T'_g of the frozen solution.

Therefore, it is important to note that the higher value of T'_g observed in frozen solutions with inulin, allowed higher freezing temperatures during processing reducing production costs.

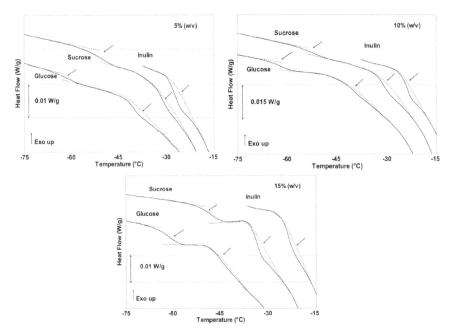

Figure 2. DSC thermograms for freeze bovine plasma protein-saccharide solutions. Down-arrows indicate T'_g. Scan rate = 2°C/min ; pH = 8.

Saccharide	Concentration (%, w/v)	T'_{g1} (°C)	T'_{g2} (°C)	T_g (°C)
Glucose	5	-62.50 ± 0.58[a]	-39.24 ± 0.75[a]	16.31 ± 0.38[a]
	10	-61.06 ± 0.45[a,b]	-39.91 ± 0.83[a]	41.52 ± 0.29[b]
	15	-59.82 ± 0.68[b]	-44.96 ± 0.49[b]	60.31 ± 0.48[c]
Sucrose	5	-51.48 ± 1.05[c]	-31.15 ± 0.40[c]	48.01 ± 0.56[d]
	10	-50.12 ± 1.03[c,d]	-31.86 ± 0.60[c]	52.48 ± 0.52[e]
	15	-48.42 ± 0.98[d]	-33.72 ± 0.45[d]	64.28 ± 0.46[f]
Inulin	5	-26.96 ± 0.68[e]	-	48.85 ± 0.35[d]
	10	-23.67 ± 0.55[f]	-	66.18 ± 0.69[g]
	15	-22.40 ± 0.45[f]	-	69.25 ± 0.45[h]

Table 1. Effect of type and concentration of cryoprotectant on glass transition temperature (T'_g) and lyoprotectant on glass transition temperature (T_g) of freeze bovine plasma proteins solutions (heating rate: 2 °C/min). Values represents the means ± standard deviation; n = 3. Values followed by different letters in the same column are significantly different from each other ($P < 0.05$).

The effect of water as a plasticizer of the mixture protein-saccharide was predicted by the Miller/Fox and Gordon–Taylor equations, the results, were compared with experimental values (Table 1). The data of T'_g of all pure components required for the Eq. (1) are listed in Table 2.

The densities of bovine plasma proteins, glucose, sucrose and inulin (at room temperature) were determined with a digital densimeter, and the results were: 0.4 ± 0.08 g/cm^3, 0.6 ± 0.05 g/cm^3, 0.8 ± 0.04 g/cm^3 and 0.3 ± 0.05 g/cm^3, respectively.

From literature the T_g of the water is -135 °C [41] and the T'_g of plasma protein is -11 ± 2 °C [22]. The T_g value of bovine plasma protein for Eq. (3), was 65 ± 3 °C. Entering this data into Eqs. (2) and (4), the predicted values of T'_g were obtained, which are listed in Table 3. The results showed that the glass transition property evaluated from the proposed models was in agreement with the experimental data with an average error of 4.86% for the Miller/ Fox equation and 0.09% for Gordon/Taylor equation. The value of k from the Gordon/Taylor equation is defined as the resistance to a T'_g decrease induced by the plasticizing effect of water [26,41,47]. The order found for k value of the saccharides was: inulin > sucrose > glucose. Although the highest value of k is for inulin, this saccharide has the highest T_g value, allowing a greater value T'_g and therefore generating a lower cost during processing, preventing also the collapse of the product at temperatures relatively higher during the freeze-drying.

Saccharide	T'_g (°C)		
	5 % (w/v)	10% (w/v)	15 % (w/v)
Glucose	-85	-79	-72
Sucrose	-59	-53	-46
Inulin	-17	-15	-13

Table 2. Data from references used in the calculation of T'_g by Miller/Fox and Gordon/Taylor modified equation [18].

	Glucose %(w/v)			Sucrose %(w/v)			Inulin %(w/v)		
	5	10	15	5	10	15	5	10	15
*ρ (g cm^{-3})	1.039	1.042	1.059	1.033	1.041	1.056	1.032	1.039	1.049
T_g' (°C) (Miller/Fox)	-63.99	-60.46	-56.1	-54.07	-51.24	-47.04	-30.43	-23.64	-19.51
Difference (%)	2.32	0.99	6.63	4.79	2.18	2.95	11.40	0.12	14.8
T_g' (°C) (Gordon/ Taylor modified)	-62.69	-61.26	-60.03	-51.38	-50.58	-48.47	-26.70	-24.23	-22.11
Difference (%)	0.30	0.33	0.35	0.19	0.91	0.10	0.97	2.31	1.31
k	3.5			4.1			4.5		

Table 3. Glass transition parameters for the multicomponent system: plasma bovine proteins-saccharides-water. *ρ: solution density (T=19.8°C)

4.2. Effect of saccharides on glass transition of the freeze-dried samples

The storage temperature of frozen or freeze-dried foods should be below the glass transition temperature as previously established [22,27,42,50]. Figure 3 shows the thermograms of the freeze-dried samples containing inulin compared with glucose and sucrose at different concentrations. The existence of these transitions evidenced the glassy state of the freeze–dried plasma protein/saccharides mixtures. Besides, Table 1 shows that T_g of the sample increases with increasing saccharide concentration. Similar results were found in the references [28,30]. This effect can be explained considering that sugar forms hydrogen–bridge bonds with proteins reducing the available volume for the interaction with water molecules, so water become less effective as plasticizer with an increase in saccharide content [51]. Also was observed that T_g of the freeze-dried samples increased with increasing of the molecular weight of the cryoprotectant. Processes of devitrification and hence product spoilage can occur if the temperature of storage is higher than the T_g of the sample. Therefore, the higher T_g value of inulin provides greater stability at higher temperatures, reducing the storage costs.

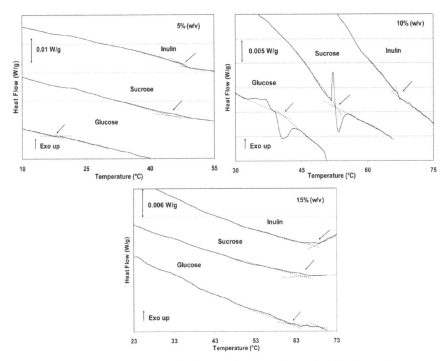

Figure 3. DSC thermograms for freeze–dried bovine plasma protein–saccharide mixtures. Down-arrows indicate T_g. Heating rate: 2°C min^{-1}; pH=8

4.3. Effect of saccharides on crystallization temperature of the freeze-dried samples

It is important to determine the crystallization temperature (T_c) of the freeze-dried samples since crystallization causes the most drastic changes on physical properties of food polymers and affects considerably food stability. The glass transition is often followed by crystallization of the solutes where the molecular mobility increases and the sample crystallizes increasing the rate of food spoilage [27,28,30].

Fig 4 shows the crystallization temperature (T_c) obtained from the intersection of the baseline and the tangent of the exothermic peak, and the crystallization enthalpy (ΔH_c) estimated as the area under the peak for the different protective agents at different concentrations. The crystallization temperature of freeze-dried samples was found to depend on the molecular weight and the saccharide concentration [27,30]. Therefore, the results showed that the presence of inulin at the same concentration than the other saccharides further increases the T_c value of freeze–dried solutions. Mixtures containing a saccharide concentration of 10 % (w/v) show an increase of ΔH_c, indicating a higher amorphous content. This behavior can be explained considering that a suitable proportion of saccharide and protein in the mixture allows a better interaction among these components [51,52,43].

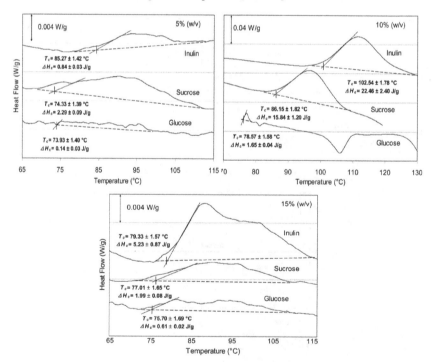

Figure 4. DSC thermogram for freeze-dried bovine plasma protein with the protective agents at different concentrations. The exothermic event indicates T_c. Heating rate: 2°C min^{-1}; pH=8.

4.4. Thermal denaturation of BPP in a matrix of saccharide

4.4.1. Effect of saccharide type and concentration

The thermal stability of BPP in a matrix of inulin compared with other saccharides was investigated using DSC. Table 4 shows the values of T_d obtained for BPP concentrate without protective agents and in different matrixes of glucose, sucrose and inulin at different concentrations. The value of T_d for BPP concentrate (88.19 ± 1.87 °C), was obtained from thermograms without protective agent and was similar to that reported in reference [53], for blood plasma. Comparing this value with the protein sample immersed in a matrix of saccharide, it was observed an increase in the value of T_d in all the cases, indicating a higher thermal resistance due to the stabilizing effect of saccharides. A similar behavior was observed in the DSC study of whey protein concentrates with the addition of honey [54]. Evaluating among the saccharides at the same concentration, it can be concluded that the higher the molecular weight of the carbohydrate, the higher was the T_d, thus inulin > sucrose > glucose. This behavior was in agreement with that reported in [55], in multi-block copolymers. With respect to the range of saccharide concentrations studied, optimum concentration was 10% (w/v), as it is shown in Table 4, in terms of the values of T_d and ΔH.

Saccharide	Concentration (%, w/v)	T_d (°C)	ΔH (J g^{-1})
Glucose		110.07 ± 1.22[a]	0.84 ± 0.32[a]
Sucrose	5	132.78 ± 2.12[b]	5.08 ± 0.98[b,c]
Inulin		143.81 ± 0.89[c]	2.97 ± 0.55[a,d,c]
Glucose		107.27 ± 0.85[a,d]	12.26 ± 0.92[e]
Sucrose	10	144.95 ± 2.34[c]	22.40 ± 1.23
Inulin		156.21 ± 1.12[e]	12.22 ± 1.43[e]
Glucose		104.91 ± 0.89[d]	3.77 ± 0.98[d,f]
Sucrose	15	126.66 ± 1.54[f]	7.01 ± 1.22[b]
Inulin		132.57 ± 1.34[b]	5.78 ± 0.76[b,f,c]

Table 4. Effect of saccharide concentration on the denaturation temperature of freeze dried BPP concentrate. Heating rate: 2°C min^{-1}. pH=8. Values followed by different letters in the same column are significantly different from each other ($P < 0.05$).

The functional structure of a protein in solution is determined by electrostatic forces, hydrogen bonds, Van der Waals interactions and hydrophobic interactions. All these interactions are influenced by water, becoming essential for the functional unfolding of most of the proteins. As water is eliminated during freeze-drying, peptide-peptide interactions prevail causing an alteration in the secondary, tertiary or quaternary structure of the protein, i.e. a conformational change of it. However, the presence of sugar displaces and supplants water forming hydrogen bonds with the dry protein which maintains its structured integrity into the glass matrix. In the case that the formation of the glass structure did not occur, the sugar would be excluded and it would not be available for the formation of hydrogen bonds to protect the dry protein from its unfolding or loss of conformation [13,14].

The protective effect of saccharides depends on its concentration, since as the concentration increases there are more possibilities of forming hydrogen bonds with the protein [11,18]. However, when concentrations were higher than 10 % (w/v), a lower protection was obtained. This result can be explained taking into account that at high concentrations, the saccharide starts to crystallize during freeze-drying, being prevented the formation of hydrogen bonds with the dry protein [12]. This behavior was confirmed by determination of the native proteins in the protein-saccharide matrixes employing eq. (1). The results are presented in Figure 5, which shows that there is a maximum at a concentration of 10% (w/v) for the different saccharides analyzed, indicating higher protein protection and stability.

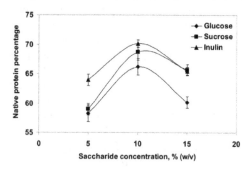

Figure 5. Native protein percentage of freeze dried BPP concentrate with different protective agents at different concentrations.

4.4.2. Effect of pH

To determine the application of these formulations is important to know the variation of T_d as a function of pH due to the wide range of environmental conditions existing in food. Table 5 shows the T_d values of BPP concentrate in a glassy matrix of saccharides at different pH values.

Saccharide	pH	T_d (°C)	ΔH (J g^{-1})
Glucose		107.27 ± 0.85[a]	12.26 ± 0.82[a]
Sucrose	8	144.95 ± 1.34[b]	22.40 ± 0.97[b]
Inulin		156.21 ± 1.12[c]	12.22 ± 0.55[a]
Glucose		102.94 ± 1.33[d]	34.74 ± 0.92[c]
Sucrose	6	134.56 ± 2.16[e]	43.15 ± 1.23[d]
Inulin		152.98 ± 1.52[c,f]	42.95 ± 1.45[d]
Glucose		101.74 ± 1.27[d]	9.58 ± 0.98[a,e]
Sucrose	4	107.67 ± 1.56[a]	9.32 ± 0.72[e]
Inulin		151.84 ± 1.89[f]	9.35 ± 0.96[e]

Table 5. Effect of pH and addition of saccharides on the denaturation temperature of BPP concentrate. Heating rate: 2°C/min. Values represents the means ± standard deviation; n = 3. Values followed by different letters in the same column are significantly different from each other (P< 0.05).

With increasing alkalinity of the medium there is an increase in the values of T_d for each saccharide (pH 8), indicating that BPP concentrate was more stable at higher pH. Similar results were found in previous works in porcine blood plasma proteins and whey protein concentrate [34,54]. Comparing between different saccharides at the same concentration, it can be seen that inulin presents a higher T_d in all the pH range. The maximum ΔH values were observed at pH 6 indicating a higher amount of native protein. Similar ΔH values at pH = 6 were reported by Dàvila in reference [34]. The lowest values of T_d and ΔH were found at pH 4, this may be to the proximity with the isoelectric point of proteins (pI: 4.8-5.8), thus decreasing the electrical net charge and facilitating aggregation reactions.

4.4.3. Effect of scanning rate

The protein-saccharide mixtures were studied at different scanning rates (2 °C/min and 5 °C/min). As an example Figure 6 shows the transition temperature and enthalpy for sucrose at 10 % (w/v).

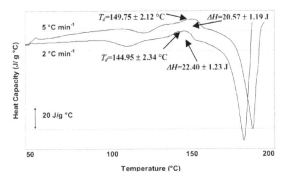

Figure 6. Effect of DSC heating rate on T_d values of freeze-dried BBP with sucrose 10%(w/v).

It was found for all the saccharides that T_d and ΔH are scanning rate dependent. T_d values increased 5 ± 2 °C in all the samples with increasing scanning rate, similar behavior was reported in references [56-58]. Furthermore, the ΔH decreased (~ 10%) with increasing scanning rate that was in agreement with the results reported in references [21,59]. Thus, the system was scanning rate dependent and so the thermal denaturation process was under kinetic control [33,44].

4.4.4. Study of Irreversibility of the Thermal Denaturation of BPP

The irreversibility of BPP denaturation was investigated by a multiple reheating experiment, according to the method proposed by by Idakieva and Michnik [45,60]. From the initial DSC scan, we have determined the values of the transition temperatures at 107°C, 145 °C and 156 °C for glucose, sucrose and inulin at 10% w/v, respectively (Table 5). DSC tests were carried out as successive scans, where the heating was carried out up to different final temperatures, with a cooling up to 20°C between scans (Figure 7).

For glucose, sucrose and inulin, the first heating was carried out up to 75°C, and 85°C (temperatures below the T_d for all the saccharides), respectively; no thermal effect was observed in the thermal denaturation peak during the reheating experiment. However, if the rescanning was stopped over their transition temperatures, the endothermic peak of T_d disappeared completely. Therefore, the endothermic peak of T_d disappeared completely upon rescanning the sample at temperatures above T_d; furthermore, as was previously described, the thermograms were scanning-rate dependent, suggesting both results that it was an irreversible event [61]. Similar behavior was also found for whey protein in an amorphous carbohydrate matrix [49], porcine blood plasma proteins [34] and BSA [33]. Irreversible denaturation of bovine plasma proteins might be due to processes such as aggregation, where hydrophobic interactions occur, and exposed thiol groups can form disulfide bonds, which result in an irreversible behavior [33]. Considering the Arrhenius law and the treatment developed in reference [43], the determination of the activation energy can be achieved from the experimental data. The obtained values were: 10443 J mol⁻¹, for BPP without protective agent; 27216 J mol-1, 32058 J mol⁻¹ and 42099 J mol⁻¹ for BPP with glucose, sucrose and inulin, respectively, all of them at 10% (w/v). The results showed that

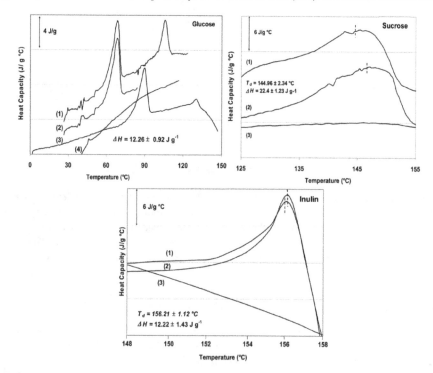

Figure 7. DSC thermograms of freeze dried BPP concentrate with saccharide at 10 % (w/v). DSC scans (2), (3), (4) represent thermograms from repeated heating and subsequent cooling. Scan (1) is a full scan to 120 °C (glucose), 155°C (sucrose) and 158 °C (inulin).

with the addition of protective agents the activation energy increased; besides with increasing molecular weight, the activation energy also increased. Therefore, the addition of saccharides, especially of inulin caused a decrease in the rate of degradation reactions, obtaining a higher stabilization upon storage [8,14,18].

4.4.5. Study of the blends morphology through SEM

Figure 8 sowed the SEM micrographs of blends of protein-saccharides.

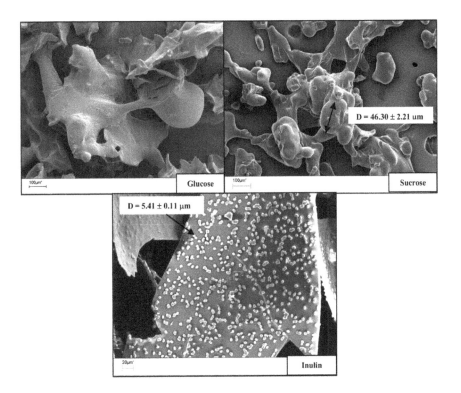

Figure 8. Scanning electron micrographs of the freeze-dried product with different saccharides, with a magnification of 200X for glucose and sucrose, 300X for inulin.

It was observed phases homogeneously distributed, indicating miscibility of the component in the matrix. The shapes were uniform, which was an attribute, linked with thermodynamic compatibility [62]. Based on the data previously obtained, comparing the transitions of the blends with respect to the value of the individual components, showed an increase in the value T_d. This increase in the T_d values can be attributed to greater miscibility of the components of the mixtures, confirming what was observed in the micrographs [61]. Therefore, these results are in agreement with the concept of miscibility, which is based on the variation of the thermal behavior with respect to the individual materials [63].

5. Conclusions

The thermodynamic properties of the solution and the freeze–dried bovine plasma proteins–saccharides mixtures were investigated in this study. The DSC thermograms demonstrated that the bovine plasma proteins– inulin mixtures have the highest glass transition temperature for the protein solution and also the highest glass transition and denaturation temperature for the freeze–dried powder, optimizing the freeze–drying process and also stabilizing and protecting the proteins during storage in conditions below the collapse temperature of the material. Thermograms revealed the existence of two glass transitions in solutions (T'_{g1} and T'_{g2}) for glucose and sucrose. With increasing saccharide content, the T'_{g1} and T'_{g2} of the samples increased and decreased, respectively. For inulin only one T'_g was found, which increased with saccharide concentration. Also was found that T'_g, T_g and T_c depended on the molecular weight of saccharides, increasing with the increasing of molecular weight, being inulin > sucrose > glucose. The proposed model allowed the prediction of transition temperature in a multicomponent mixture which is useful to design a freeze–drying cycle and storage stability of plasma protein concentrates. The addition of saccharides allowed the increase of the protein denaturation temperature and enthalpy, with an optimal saccharide concentration of 10% (w/v) and a pH range between 6 and 8. This change in the thermal properties shows a greater compatibility of the blends with 10% (w/v) saccharide, because this concentration causes the greatest changes in the values of T_d when compared with individual values of BPP. The results were corroborated by the SEM micrographs, showing homogeneously distributed phases, and denoting the highest miscibility between them. The temperature of thermal denaturation was scan rate dependent, and no thermal transition was detected in the re-scan experiments so it was concluded that the protein unfolding was irreversible and was adequately interpreted by the theoretical model employed.

Therefore, the results showed highest values of T_g and T_d in the freeze–dried samples of inulin proven that this compound is a better protein protective agent during storage than mono and disaccharides such as glucose and sucrose. In this way prevent the unfolding of bovine plasma proteins submitted to higher temperatures. Furthermore, the higher T'_g of frozen solutions of bovine proteins with inulin allows higher freezer temperatures during freeze–drying, reducing costs in a food elaboration. The finding about the inulin cryoprotant role of food proteins is relevant considering that it is a soluble fiber, categorized as a prebiotic, and being a valuable alternative as a functional ingredient for food formulation [64,65].

The findings regarding the protective effect of inulin on bovine plasma proteins, suggest that may be interesting the study of the behavior of formulated foods elaborated with the analyzed matrices (protein-saccharide-water) exposed to treatments such as cooling and freeze-drying.

Author details

Laura T. Rodriguez Furlán, Antonio Pérez Padilla and Mercedes E. Campderrós*
Research Institute of Chemical Technology (INTEQUI –CONICET-CCT San Luis), Faculty of Chemistry, Biochemistry and Pharmacy, UNSL, San Luis, Argentine

Javier Lecot and Noemi E. Zaritzky
Research and Development in Food Cryotechnology Centre (CIDCA- CONICET- CCT La Plata), Argentine

Noemi E. Zaritzky
Faculty of Engineering, UNLP, La Plata, Bs As, Argentine

6. References

[1] Niness K R (1999) Inulin and Oligofructose: What Are They? J. nutr. 129 (7): 1402-1406.
[2] Abrams S, Griffin I, Hawthorne K, Liang L, Gunn S, Darlington G, Ellis K A (2005) Combination of Prebiotic Short- and Long-chain Inulin-type Fructans Enhances Calcium Absorption and Bone Mineralization in Young Adolescents. Am. j. clin. nutr. 82: 471-476.
[3] Hempel S, Jacob A, Rohm H (2007) Influence of Inulin Modification and Flour Type on the Sensory Quality of Prebiotic Wafer Crackers. Eur. food res. technol. 224: 335-341.
[4] Nazzaro F, Fratianni F, Coppola R, Sada A, Pierangelo O (2009) Fermentative Ability of Alginate-prebiotic Encapsulated Lactobacillus Acidophilus and Survival under Simulated Gastrointestinal Conditions. J. funct. food 1(3): 319-323.
[5] Kip P, Meyer D, Jellema R H (2006) Inulins Improve Sensory and Textural Properties of Low-Fat Yoghurts. Int. dairy j. 16: 1098–1103.
[6] Ronkart S N, Paquot M, Fougnies C, Deroanne C, Blecker C S (2009) Effect of Water Uptake on Amorphous Inulin Properties. Food hydrocolloid 23: 922–927.
[7] Baeza R I, Pilosof A M R (2002) Calorimetric Studies of Thermal Denaturation of b-Lactoglobulin in the Presence of Polysaccharides. Lebensm.-wiss. technol. 35: 393–399.
[8] Buera P, Schebor C, Elizalde B (2005) Effects of Carbohydrate Crystallization on Stability of Dehydrated Foods and Ingredient Formulations. J. food eng. 67: 157-165.
[9] Claude J, Ubbink J (2006) Thermal Degradation of Carbohydrate Polymers in Amorphous States: A Physical Study Including Colorimetry. Food chem. 96: 402-410.
[10] Santivarangkna C, Higl B, Foerst P (2008) Protection Mechanisms of Sugars During Different Stages of Preparation Process of Dried Lactic Acid Starter Cultures. Food microbiol. 25: 429-441.
[11] Allison S D, Chang B, Randolph T W, Carpenter J F (1999) Hydrogen Bonding Between Sugar and Protein is Responsible for Inhibition of Dehydration-Induced Protein Unfolding. Biochem. biophys. 365: 289-298.
[12] Carpenter J F, Crowe L M, Crowe J H (1987) Stabilization of Phosphofructokinase with Sugars during Freeze-drying: Characterization of Enhanced Protection in the Presence of Divalent Cations. Biochim. biophys. acta 923(1): 109-115.

---

* Corresponding Author

[13] Hinrichs W L J, Prinsen M G, Frijlink H W (2001) Inulin Glasses for the Stabilization of Therapeutic Proteins. Int. j. pharmaceut. 215: 163–174.

[14] Hinrichs W L J, Sanders N N, De Smedt S C, Demeester J, Frijlink H W (2005) Inulin is a Promising Cryo- and Lyoprotectant for PEGylated Lipoplexes. J. control. release 103: 465-479.

[15] Rey Cabinet L, d'Etudes L, Switzerland J C (2004) Freeze Drying/lyophilization of Pharmaceutical and Biological Products. Maryland, U.S.A: Center for Biologics Evaluation and Research Food and Drug Administration.

[16] Liao Y H, Brown M B, Martin G P (2004) Investigation of the Stabilization of Freeze–dried Lysozyme and the Physical Properties of the Formulations. Eur j. pharm. Biopharm. 58: 15–24.

[17] Minson E I, Fennema O, Amundson C H (2006) Efficacy of Various Carbohydrates as Cryoprotectants for Casein in Skim Milk. J. food sci. 46(5): 1597-1602.

[18] Rodriguez Furlán L T, Pérez Padilla A, Campderrós M (2010) Inulin Like Lyoprotectant of Bovine Plasma Proteins Concentrated by Ultrafiltration. Food res. int. 43: 788-796.

[19] Costantino H R, Curley J G, Wu S, Hsu C C (1998) Water Sorption Behavior of Lyophilized Protein–sugar Systems and Implications for Solid-state Interactions. Int. j. pharm. 166: 211–221.

[20] Passot S, Fonseca F, Alarcon-Lorca M, Rolland D, Marin M (2005) Physical Characterization of Formulations for the Development of Two Stable Freeze–dried Proteins During Both Dried and Liquid Storage. Eur. j. pharm. biopharm. 60: 335–348.

[21] Guzzi R, Sportelli L, Sato K, Cannistraro S, Dennison C (2008) Thermal Unfolding Studies of a Phytocyanin. Biochim. biophys. acta 1784: 1997-2003.

[22] Chen T, Oakley D M (1995) Thermal Analysis of Proteins of Pharmaceutical Interest. Thermochim. acta 248: 229–244.

[23] Schenz T W (1995) Glass Transitions and Product Stability–an Overview. Food hydrocolloid 9(4): 307–315.

[24] Ahmed J, Prabhu S T, Raghavan G S V, Ngadi M. (2007) Physico-chemical, Rheological, Calorimetric and Dielectric Behavior of Selected Indian Honey. J. food eng. 79: 1207-1213.

[25] Gallegos Infante J A, Ochoa Martínez L A, Ortiz Corral C (2005) Glass Transition Temperature Behavior of a Model Blend of Carbohydrates. Cien. Tecnolog. Alimen. 5: 6–10.

[26] Noel T R, Parker R, Ring S G, Tatham A S (1995) The Glass-transition Behaviour of Wheat Gluten Proteins. Int j. boil. macromol. 17 (2): 81–85.

[27] Roos Y (1995) Characterization of Food Polymers using State Diagrams. J. food eng. 24: 339–360.

[28] Shah B N, Schall C A (2006) Measurement and Modeling of the Glass Transition Temperatures of Multi-component Solutions. Thermochim. acta 443: 78– 86.

[29] Katkov I I, Levine F (2004) Prediction of the Glass Transition Temperature of Water Solutions: Comparison of Different Models. Cryobiology 49: 62–82.

[30] Tattini Jr V, Parra D F, Polakiewicz B, Pitombo R N M (2005) Effect of Lyophilization on the Structure and Phase Changes of PEGylated-bovine Serum Albumin. Int. j. pharm. 304: 124–134.

[31] Sunooj K V, Radhakrishna K, George J, Bawa A S (2009) Factors Influencing the Calorimetric Determination of Glass Transition Temperature in Foods: a Case Study Using Chicken and Mutton. J. food eng. 91: 347–352.

[32] Akköse A, Aktas N (2008) Determination of Glass Transition Temperature of Beef and Effects of Various Cryoprotective Agents on Some Chemical Changes. Meat sci. 80: 875–878.

[33] Cao X, Li J, Yang X, Duan Y, Liu Y, Wang C (2008) Nonisothermal Kinetic Analysis of the Effect of Protein Concentration on BSA Aggregation at High Concentration by DSC. Thermochim. acta 467: 99-106.

[34] Dàvila E, Parés D, Cuvelier G, Relkin P (2007) Heat-induced Gelation of Porcine Blood Plasma Proteins as Affected by pH. Meat Sci. 76: 216-225.

[35] de Wit J N (1990) Thermal Stability and Functionality of Whey Proteins. J. dairy Sci. 73: 3602-3612.

[36] Giroux H J, Britten M (2004) Heat Treatment of Whey Proteins in the Presence of Anionic Surfactants. Food hydrocolloid 18: 685- 692.

[37] SAS USER GUIDE: Statistic. Versión (1989). SAS Inst. Inc., Cary, NC, USA.

[38] Fox T G (1956) Influence of Diluent and Copolymer Composition on the Glass Temperature of a Polymer System. B. am. phys. soc. 2(1): 123.

[39] Miller D P, de Pablo J J, Corti H (1997) Thermophysical Properties of Trehalose and its Concentrated Aqueous Solutions. Pharm res-dord 14(5): 578–590.

[40] Gordon M, Taylor J S (1952) Ideal Copolymers and the Second-order Transitions of Synthetic Rubbers. I. Non-crystalline Copolymers. J. appl. chem. 2: 493–500.

[41] Georget D M R, Smith A C, Waldron K W (1999) Thermal Transitions in Freeze– dried Carrot and its Cell Wall Components. Thermochim. acta 332: 203–210.

[42] Maitani Y, Aso Y, Yamada A, Yoshioka S (2008) Effect of Sugars on Storage Stability of Lyophilized Liposome/DNA Complexes with High Transfection Efficiency. Int. j. pharm. 356: 69–75.

[43] Rodriguez Furlán L T, Lecot J, Pérez Padilla A, Campderrós M, Zaritzky N (2011) Effect of Saccharides on Glass Transition Temperatures of Frozen and Freeze-dried Bovine Plasma Protein. J. food eng. 106: 74-79.

[44] Creveld L D, Meijberg W, Berendsen H J C, Pepermans H A M (2001) DSC Studies of Fusarium Solani Pisi Cutinase: Consequences for Stability in the Presence of Surfactants. Biophys. Chem. 92: 65-75.

[45] Idakieva K, Parvanova K, Todinova S (2005) Differential Scanning Calorimetry of the Irreversible Denaturation of Rapana Thomasiana (Marine Snail, Gastropod) Hemocyanin. Biochim. biophys. acta 1748: 50-56.

[46] Ramprakash J, Doseeva V, Galkin A, Krajewski W, Muthukumar L, Pullalarevu S, Demirkan E, Herzberg O, Moult J, Schwarz, F P (2008) Comparison of the Chemical and Thermal Denaturation of Proteins by a Two-state Transition Model. Anal. biochem. 374: 221-230.

[47] Ohkuma C, Kawaib K, Viriyarattanasaka C, Mahawanichc T, Tantratianc S, Takaia R, Suzuki T (2008) Glass Transition Properties of Frozen and Freeze–dried Surimi Products: Effects of Sugar and Moisture on the Glass Transition. Food hydrocolloid 22: 255–262.

[48] Telis V R N, Sobral P J A (2002) Glass Transitions for Freeze–dried and Air–dried Tomato. Food res. int. 35: 435–443.

[49] Sun W Q, Davidson P, Chan H S O (1998) Protein Stability in the Amorphous Carbohydrate Matrix: Relevance to Anhydrobiosis. Biochim. biophys. acta 1425: 245-254.

[50] Salman A D, Hounslow M J, Seville J P K (2006) Granulation. In: Sal, A. (Ed.), Handbook of Powder Technology, vol. 11. España: Elsevier.

[51] Gabbott P (2008) Principles and Applications of Thermal Analysis. Blackwell Publishing. Chapter 9.

[52] Dilworth S E, Buckton G, Gaisford S, Ramos R (2004) Approaches to Determine the Enthalpy of Crystallization, and Amorphous Content, of Lactose from Isothermal Calorimetric Data. Int. j. pharm. 284: 83–94.

[53] [53] Relkin P (1996) Thermal Unfolding of β-Lactoglobulin, α-Lactalbumin and Bovine Serum Albumin. A Thermodynamic Approach. Crit. rev. food sci. 36 (6): 565–601.

[54] Yamul D K, Lupano C E (2003) Properties of Gels from Whey Protein Concentrate and Honey at Different pHs. Food res. int. 36: 25-33.

[55] Penco M, Sartore L, Bignotti F, D'Antone S, Di Landro L. (2000) Thermal Properties of a New Class of Block Copolymers Based on Segments of Poly(D,L-lacticglycolic Acid) and Poly(e-caprolactone). Eur. polym. j. 36: 901-908.

[56] Kavitha M, Bobbili K B, Swamy M J (2010) Differential Scanning Calorimetric and Spectroscopic Studies on the Unfolding of Momordica Charantia Lectin. Similar Modes of Thermal and Chemical Denaturation. Biochimie 92: 58-64.

[57] Schubring R (1999) DSC Studies on Deep Frozen Fishery Products. Thermochim. acta 337: 89-95.

[58] Zamorano L S, Pina D G, Gavilanes F, Roig M G, Yu Sakharov I, Jadan A P, van Huystee, R B, Villar E, Shnyrov V L (2004) Two-state Irreversible Thermal Denaturation of Anionic Peanut (Arachis Hypogaea L.) Peroxidase. Thermochim. acta 417: 67-73.

[59] Vermeer A W P, Norde W (2000) The Thermal Stability of Immunoglobulin: Unfolding and Aggregation of a Multi-Domain Protein. Biophys. j. 78: 394-404.

[60] Michnik A, Drzazga Z, Kluczewska A, Michalik K (2005) Differential Scanning Microcalorimetry Study of the Thermal Denaturation of Haemoglobin. Biophys. chem. 118: 93-101.

[61] Rodriguez Furlán L T, Lecot J, Pérez Padilla A, Campderrós M E, Zaritzky N (2012) Stabilizing Effect of Saccharides on Bovine Plasma Protein: A Calorimetric Study". Meat sci. In press.

[62] Gallego K, López B L, Gartner C (2006) Estudio de Mezclas de Polímeros Reciclados para el Mejoramiento de sus Propiedades. Rev. Fac. Ing. 37: 59-70.

[63] Mousavioun P, Doherty W O S, George G (2010) Thermal Stability and Miscibility of Poly(hydroxybutyrate) and Soda Lignin Blends. Ind. crop. prod. 32(3): 656-661.

[64] Rodriguez Furlán L T, Pérez Padilla A, Campderrós M E (2010) Functional and Physical Properties of Bovine Plasma Proteins as a Function of Processing and pH, Application in a Food Formulation. Adv. j. food sci. tech. 2(5): 256-267.

[65] Rodriguez Furlán L T, Rinaldoni A N, Padilla A P, Campderrós M E (2011) Assessment of Functional Properties of Bovine Plasma Proteins Compared with other Proteins Concentrates, Application in a Hamburger Formulation. Am. j. food tech. 6 (9): 717-729.

Chapter 9

Determination of Folding Reversibility of Lysozyme Crystals Using Microcalorimetry

Amal A. Elkordy, Robert T. Forbes and Brian W. Barry

Additional information is available at the end of the chapter

1. Introduction

An important aspect in the preparation of proteins as pharmaceutical products is stabilisation of the native protein conformation (folded, three-dimensional, tertiary state), which is required for biological activity. Moreover, it is not enough for this conformation to be stable, but the protein must be able to find the state or folding pathway in a short time from a denatured, unfolded conformation [1]. Folding minimises exposure of non-polar groups and maximises exposure of polar groups to the solvent [2].

Lysozyme, a globular protein, molecular weight 14,300 Da, was chosen as a model protein; it consists of a single 129 amino acid chain divided into two domains cross linked by four disulfide bridges. The hydrophilic groups tend to concentrate on the surface and the hydrophobic groups in the core [3]. The goal of this study was to investigate the influences of crystallisation on folding reversibility of lysozyme in solution as assessed calorimetrically.

Many literature reports cited the value of High Sensitivity Differential Scanning Calorimetry (HSDSC) for determining thermodynamic parameters (transition temperature, T_m, and enthalpies, ΔH) that describe the folded and unfolded states [4-6]. Furthermore, HSDSC was used to measure thermal transition reversibility that is no less important than T_m and ΔH [7-8]; a protein transition is considered reversible if the molecule renatures upon cooling after heat treatment. Thermodynamic or conformational stability is defined as the difference in free energy (ΔG) between the folded and unfolded state. This stability is the sum of weak non-covalent interactions including hydrogen bonds, van der Waal interactions, salt bridges and hydrophobic forces [9]. Thermodynamic stability is divided into biophysical, which includes the study of thermodynamics, protein denaturation and renaturation (as is discussed in this Chapter) and biochemical, which involves comparative studies of protein conformation and stability of two or more proteins to establish various structural features that deduce stability within a given biomolecule [10].

Differential scanning calorimetry has the ability to provide detailed information about both the physical and energetic properties of a substance [11]. The HSDSC technique was described in detail by Cooper and Johnson (1994) [12]. In summary, all equilibrium processes involving molecules are governed by free energy changes (ΔG), made up of enthalpy (ΔH) and entropy (ΔS). The relationship between (ΔH) and (ΔS) is given by the Gibbs free energy equation:

$$\Delta G = \Delta H - T.\Delta S \tag{1}$$

where T is the temperature. Differential scanning calorimeters measure enthalpies, which provide the basis for determining the thermodynamic properties of a system. Both enthalpy and entropy are related to the heat capacity of the system. The enthalpy is the total energy (at constant pressure) required to heat the system from absolute zero to the required temperature:

$$H = \int_0^T C_P(T).dT \tag{2}$$

where C_P (T) is the temperature-dependent heat capacity at constant pressure. The total entropy of the system can be expressed as:

$$S = \int_0^T [C_P(T)/T].dT \tag{3}$$

Accordingly, differences in H and S can be expressed as:

$$\Delta H = \int_0^T \Delta C_P(T).dT \tag{4}$$

and

$$\Delta S = \int_0^T \Delta [C_P(T)/T].dT \tag{5}$$

Thermodynamic parameters depend on conditions, such as temperature, pressure, concentration and composition. Thus, it is necessary to correct experimental results to standard conditions (standard states) denoted by the superscript 0 (e.g. ΔG^0) for simplicity of comparison of data from different situations. The standard state of solutes in dilute solutions is a concentration of 1 M. The standard temperature and pressure are usually 25°C and 1 atm, respectively. The standard free energy change (ΔG^0), which is the free energy change during the reaction where all components are in their standard states, can be measured from:

$$\Delta G^0 = - RT.\ln K \tag{6}$$

where R is the gas constant and K is the equilibrium constant for the process and is related to (ΔH^0) and (ΔS^0) and to the absolute temperature by:

$$\ln K = - (\Delta H^0/RT) + (\Delta S^0/R) \tag{7}$$

assuming that (ΔH^0) and (ΔS^0) do not vary significantly with temperature over the range of interest.

The HSDSC provides an insight into the thermal stability and instability (e.g. formation of soluble and insoluble aggregates) of solutions of different formulations. HSDSC was used to assess the thermal stability of lysozyme solutions after storage at stressed conditions [13]. Consecutive heating scans indicated the folding reversibility of thermal transitions [8, 14] and the validity of calorimetrically measured protein folding reversibility [15-17]. Creighton (1994) [18] reported mechanisms and thermodynamic factors controlling protein folding-and-unfolding.

In the present study, HSDSC investigated thermal changes; in particular protein refolding performance, of crystallised samples (in low and high protein concentrations) upon heat treatment. The thermal structural transition of lysozyme involves two thermodynamic states, native and denatured [19] as for other globular proteins [20]. However, Hirai et al. (1999) [21] indicated that folding-and-unfolding kinetics of proteins depend on the number of amino acid residues. Proteins with residues above ~100 do not follow simple two state kinetics in a folding-and-unfolding process as a single cooperative unit. Accordingly during HSDSC analysis of lysozyme, there might be formation of intermediates between native and denatured states. The thermodynamic stability of proteins not only requires that the transition temperature (T_m) and other thermodynamic parameters remain constant but also implies reversibility of protein from unfolded (denatured) to folded (native) state after removing the effect of an external condition such as heat. Anfinsen (1973) [22] reported that denaturation of Ribonuclease A, by heat or urea, was reversible when denatured molecules returned to a normal environment of temperature and solvent. Hence, both structure and enzymatic activity were regained.

Consequently, formation of the native state is a global property of the protein as described [1]. This state is necessary for stability and activity; proteins are marginally stable and achieve stability only within narrow ranges of conditions of solvent and temperature. The free energy of stabilization of proteins under ordinary conditions is ~ 5-15 kcal mol^{-1} [1].

Proteins undergo various structural changes if physiological conditions alter. Accordingly, they may denature and the denatured protein tends to adsorb to surfaces and aggregate with other protein molecules. Katakam et al. (1995) [23] proposed that denaturation of recombinant human growth hormone involves unfolding of the molecule; the unfolded part adsorbs to surfaces and aggregates with neighbouring molecules. Shaking and exposure to an air/water interface, heating, lyophilisation or reconstitution of lyophilised protein may aggregate protein with subsequent loss of stability and activity.

The combination of HSDSC and enzymatic activity determined if refolding of denatured crystallised lysozyme after thermal denaturation in HSDSC arises from the nativeness, three-dimensional folded state, of the initial lysozyme structure. This means that enzymatic activity was employed to investigate if folding reversibility of the thermal transition reflects the renaturation of the unfolded protein to folded native structure.

2. Materials and methods

2.1. Materials

Chicken egg white lysozyme (purity 95%, 5% sodium chloride and sodium acetate), sodium chloride (99.5%), sodium phosphate (99.3%) and *Micrococcus lysodeikticus* were purchased from Sigma Chemical Company (St. Louis, Mo). Sodium acetate anhydrous (98%), potassium dihydrogen orthophosphate (> 99%) were obtained from BDH Chemicals Ltd. Poole, UK. Water was deionised, double distilled.

2.2. Preparation of crystallized lysozyme

Lysozyme was crystallised using a published method [24]. Crystals formed were filtered, dried and kept in a freezer (-15°C) until tested.

2.3. High Sensitivity Differential Scanning Calorimetry (HSDSC)

Solution samples of crystallised lysozyme were analysed with a Microcal MCS differential scanning calorimeter (Microcal Inc., MA, USA). Degassed samples (5 and 20 mg product / 1 mL 0.1M sodium acetate buffer, pH 4.6) and reference (0.1M sodium acetate buffer, pH 4.6) were loaded into cells using a gas tight Hamilton 2.5 mL glass syringe. The folding reversibility of lysozyme denaturation was assessed by temperature cycling using two scan calorimetric methods. The upscan-upscan method (UU) employed two consecutive upscans from 20-90°C at 1°C/min. After the first upscan, the sample was immediately cooled in the calorimeter (downscan) to 20°C at 0.75°C /min (the fastest cooling rate allowed by the instrument) and the heating cycle was immediately repeated. Transition reversibility was measured as ratio (%) of enthalpy change of second upscan (ΔH_2) over that of first upscan (ΔH_1). The upscan-downscan method (UD) involved heating of protein solution from 20-90°C at 1°C/min immediately followed by downscan (cooling) from 90-20°C at a cooling rate of 0.75°C/min. Enthalpies were measured and downscan (ΔH_3) / upscan (ΔH_1) enthalpy ratios were calculated as a measure of folding reversibility. The calorimeter was temperature- and heat capacity-calibrated using sealed hydrocarbon standards of known melting points and electrical pulses of known power, respectively.

Experiments were performed under 2 bar nitrogen pressure. A base line was run before each measurement by loading the reference in both the sample and reference cells; this base line was subtracted from the protein thermal data and the excess heat capacity was normalized for lysozyme concentration. Data analysis and deconvolution employed ORIGIN DSC data analysis software. The T_m (mid point of the transition peak) values for all transitions were calculated.

2.4. Enzymatic assay

Biological activities of thermally denatured crystallised lysozyme were determined after cooling (in HSDSC) to determine whether the renaturation is due to the nativeness of the protein structure i.e. to correlate the folding reversibility with biological activity. In this assay, a bacterial suspension was prepared by adding 20 mg of *Micrococcus lysodeikticus* to 90 mL of phosphate buffer 0.067 M, pH 6.6, and 10 mL of 1% NaCl. The biological reaction was initiated by addition of 0.5 mL of each enzyme solution to 5 mL of the bacterial suspension. The activity unit of lysozyme is defined as the amount of enzyme decreasing the absorption rate at 450 nm at 0.001 /min at 25°C and pH 6.6. Rates were monitored using a UV/Vis. spectrophotometer (Pu 8700, Philips, UK) at 25°C.

All data were presented as mean of three determinations ± standard deviation. The Student's *t*-test assessed significance.

3. Results and discussion

3.1. High Sensitivity Differential Scanning Calorimetry (HSDSC)

Differential scanning microcalorimetry experiments can thermodynamically characterise the unfolding transition by determing heat capacities, enthalpies and melting temperatures of native and denatured protein [25]. HSDSC monitored thermal stability and folding reversibility of reconstituted lysozyme preparations. For samples, traces for thermal denaturation and folding reversibilities, using (UU) method, of reconstituted crystallised lysozyme are illustrated in Figure 1(a) and (b) for 5mg/mL and 20mg/mL protein concentrations, respectively. Thermodynamic parameters and enzymatic activities are in Table 1. Figure 2 shows an example for folding reversibility of unprocessed lysozyme using (UD) method. As is evident in Figures 1 and 2, HSDSC profiles of all samples showed a single endothermic peak (first upscan). Lysozyme crystals started to unfold at ~65°C with a mean T_m of 76.1°C (T_{m1}).

It is noticeable that rescan profiles, whether endothermic (second heating cycle, Figure 1a and b dotted lines) or exothermic (downscan upon cooling, Figure 2 lower trace) showed two peaks, a main one and a small peak or shoulder. Deconvolution of the data (using ORIGIN DSC data analysis software) revealed two transition regions characterised by T_m at ~76.1°C (T_{m2}) for the main peak and at ~66°C for the shoulder, indicating that the lysozyme transition is not a two state transition. This may be explained on the basis that lysozyme consists of a single polypeptide chain divided into two structural domains (α-helix and β-sheet). During refolding each domain may be refolded separately with different pathways. Consequently, two peaks appear instead of one; this explanation agrees with Remmele et al. (1998) [14] who attributed the three T_m peaks to the three domains of immunoglobulin-type domains that make up the Interleukin-1 receptor.

The other reasonable explanation is that lysozyme, when its folding process is analysed using circular dichroism, does not obey a single co-operative transition, but the process involves several parallel folding pathways. Each of the two domains stabilises with different kinetics [26]. In particular, the amides in the α-helix are involved in the formation of stable helical structure and assembly of the hydrophobic core. Then a stable hydrogen bonded structure in the β- domain forms. Accordingly, partially structured intermediates develop during the folding of lysozyme. This explanation is supported by Buck et al. (1993) [27] who reported that lysozyme consists of two structural domains that are stabilised by different pathways.

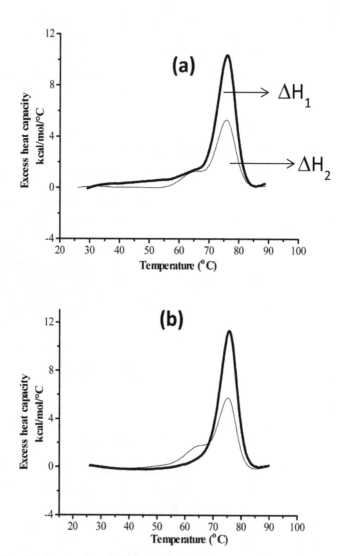

Figure 1. Normalised consecutive calorimetric upscans, first upscan (solid line) and second upscan (dotted line) of crystallised lysozyme. Conditions: (a) 5mg/mL protein, 0.1 M sodium acetate buffer, pH 4.6, heating rate 1°C/min and (b) 20mg/mL protein, 0.1 M sodium acetate buffer, pH 4.6, heating rate 1°C/min.

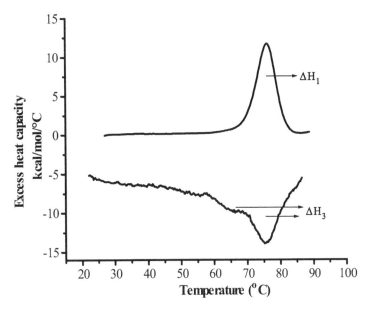

Figure 2. Normalised calorimetric upscan (upper trace) and downscan (lower trace) of lysozyme, as received. Conditions: 5mg/mL protein, 0.1 M sodium acetate buffer, pH 4.6, heating rate 1°C/min, cooling rate 0.75°C/min.

Lysozyme Sample	T_{m1} (°C)	T_{m2} (°C)	% folding reversibility ($\Delta H_2/\Delta H_1$)	% enzymatic activity
Crystallised				
5 mg mL^{-1}	76.1±0.21	75.9±0.15	66.5±1.4	65.7±1.4
20 mg mL^{-1}	75.6±0.07	75.2±0.02	52.4±1.6	52.7±2.2

[a] T_{m1}, T_{m2} are mid-point peak transition temperatures of first and second upscans; ΔH_1, ΔH_2 are calorimetric enthalpies of transitions of first and second upscans; % enzymatic activity is the activity of each sample relative to fresh material of that sample, ±S.D., n= 3.

Table 1. Thermodynamic parameters for the thermal denaturation, folding reversibilities using consecutive upscan method (UU) and enzymatic activities of crystallised lysozyme samples[a].

Lysozyme Sample	% folding reversibility ($\Delta H_3/\Delta H_1$)
Crystallised	
5 mg mL^{-1}	43.6 (1.6)
20 mg mL^{-1}	48.6 (3.8)

[a] Values between brackets are S.D., n= 3.

Table 2. Folding reversibilities using upscan-downscan (UD) method of crystallised lysozyme samples [a].

A comparison of calorimetric enthalpy (ΔH_{cal}) and the theoretical enthalpy (ΔH_{VH}, van't Hoff enthalpy) changes, to judge the validity of a two-state mechanism for the unfolding of lysozyme, reveals the presence of intermediates [25]. It was reported that, in the unfolded state, proteins aggregate and react chemically with amino acid residues exposed to the solvent; this can lead to misfolding or irreversible denaturation [28]. Also, the formation of any irreversible component alters the shape of a HSDSC thermodynamic peak over the temperature range at which it forms [29].

For low protein concentration (5mg/mL, Figure 1a), Table 1 shows no significant difference between T_{m1} (transition temperature of first upscan) and T_{m2} (transition temperature of second upscan) for protein samples.

It was reported that determination of the mechanism and pathway of unfolding and refolding depends on the identification of the intermediates that may not be stable at the equilibrium [18]. Thus, detection and characterisation of kinetic folding intermediates is complex. This intricacy can arise from accumulation of intermediates or from subpopulations of the unfolded state refolding at different rates. Also, events in folding are obscure [1]. With respect to samples with high protein concentration (20mg/mL, Figure 1b), Table 1 demonstrates that T_{m2} decreased compared to its corresponding T_{m1}.

On comparing low and high protein concentrations, thermal stabilities (T_{m1} and T_{m2}) of lysozyme crystals at high concentration significantly decreased ($p < 0.05$). Accordingly, high protein concentration influences thermal stability, this is in agreement with previously published data for dried proteins [17]. Moreover, folding reversibilities and enthalpies of first upscan of all samples (Figure 1 and Table 1) decreased with increasing concentration. Enthalpy values correlated with the content of ordered secondary structure of protein [30]. The decrease in enthalpy of protein may be attributed to denaturation at high protein concentration because a partially unfolded protein needs less heat energy to denature than a native form [17]. In general, crystals are chemically and physically pure substances (atoms or molecules within crystals are arranged in highly ordered patterns in three dimensional structures); the other possible reason for the observed reversibility of lysozyme at high protein concentration is that the water in lysozyme crystal lattices inhibits protein-protein interactions and aggregation, to some extent, which may take place at high lysozyme concentration after denaturation by heat in the HSDSC. Consequently, the crystallisation maintains the three-dimensional folded structure of lysozyme and enhances the renaturation of the protein. Water molecules play an important role in the function of proteins through maintaining their tertiary structure. The structure of biological macromolecules in an aqueous solution is similar to that in a crystalline state [31]. There are two kinds of hydration shell of biomolecules in aqueous solution; the primary and secondary hydration shells. Water molecules in the primary hydration shell are directly bound to molecules. The water molecules in the secondary hydration shell have a character intermediate between those of the primary hydration shell and bulk water. On the other hand, crystal water is classified into groups that correspond to the hydration shells in solutions. These water molecules correspond mainly to those in the primary hydration shell and partly to those in the secondary hydration shell [31]. Consequently, there is strong interaction between water and protein molecules in crystalline states. Takano et al. (1999) [6] used differential scanning calorimetry to examine the contribution of hydrogen bonds to the conformational stability of mutant human lysozyme. The authors commented that hydrogen bonding between human lysozyme atoms and water bound with the protein molecules in crystals contributes to the protein conformational stability. The net contribution of one intramolecular hydrogen bond to protein stability in terms of Gibbs energy was ~8.9 kJ/mol. On the basis of Takano et al. (1999) [6] study, hydrogen bonds are one of the important factors stabilising the folded conformations of proteins. From these results crystals were capable of refolding and hence lysozyme crystals not only maintained thermal stability and conformational integrity as suggested previously [24], but also improved refolding ability, which is necessary in protein formulation and processing. Refer to a study by Elkordy et al. [17] for folding reversibility of lyophilised and spray dried lysozyme. Also, a study by Forbes et al. [16] reported the folding reversibility of spray dried and crystallised trypsin.

For folding reversibility calculated by (UD) method, Table 2 above summarises the results of folding reversibilities of crystallised lysozyme at low and high protein concentrations. From Table 2, it is apparent that the percentage folding reversibility calculated by (UU) method (Table 1) was significantly higher ($p < 0.05$) than that derived from the (UD) method (Table 2). This implies that the latter method underestimates the apparent folding reversibility of samples.

3.2. Enzymatic assay

Lysozyme solutions upon cooling in the HSDSC after thermal denaturation were assayed for biological activity towards *Micrococcus lysodeikticus*. Based on the HSDSC results, all samples renatured to some extent after thermal stress. Thus, enzymatic assay should answer an important question. Is this renaturation or folding reversibility related to regain of the native structure of lysozyme (which is essential for biological activity), or does it result from misfolding, i.e. folding of the protein in a manner different from the original natured structure which subsequently leads to loss of activity?

Table 1 presents percentage enzymatic activities of preheated solutions, in HSDSC, of crystallised lysozyme (relative to an aqueous solution of a fresh sample). It is evident that the biological activity of lysozyme was maintained by crystals (5mg/mL and 20mg/mL). The results were consistent with data of folding reversibilities. This answers the question posed previously in that folding reversibility was related to the native structure of lysozyme that is required for its activity, as the greater the folding reversibility, the higher the enzymatic activity. The results illustrated that lysozyme crystals maintained structural integrity even after heating in the HSDSC. A review by Jen and Merkle (2001) [32] showed that hydrated protein within crystals is present in a folded, native form.

From the HSDSC and enzymatic activity results, the folding reversibility, calculated by consecutive upscans (UU, Tables 1), correlated with enzymatic activity of lysozyme, confirming that the upscan-downscan method (UD, Table 2) underestimates the magnitude of folding reversibility. However, proteins are diverse molecules and the presence of correlation between folding reversibility and biological activity of lysozyme, as demonstrated in this study, may not be applicable to other proteins.

4. Conclusions

The overall results suggested that reconstituted lysozyme crystals were able to refold after heating. The folding reversibility arises from the nativeness of the initial lysozyme structure as demonstrated by biological activity data. The results indicated that the upscan-downscan method underestimated the extent of folding reversibility. Consequently, it is preferable to calculate this reversibility, employing high sensitivity differential scanning calorimetry, by the consecutive heating upscan method.

Author details

Amal A. Elkordy*
Department of Pharmacy, Health and Well-being, University of Sunderland, Sunderland, UK

Robert T. Forbes and Brian W. Barry
School of Pharmacy, University of Bradford, Bradford, UK

5. References

[1] Lesk A M 2001. In vivo, in vitro, in silicio. In: Lesk, A.M. (Ed.), Introduction to protein architecture. Oxford University Press, Oxford, New York, pp.: 15-35, 143.

[2] Rupley J A, Gratton E, Careri G 1983. Water and globular proteins. Trends in Biochem Sci. 8: 18-22.

[3] Rosenberger F 1996. Protein crystallization. J Cryst Growth. 166: 40-54.

[4] Pikal M J, Lukes A L, Lang J E, Gaines K 1978. Quantitative crystallinity determination for b-lactam antibiotics by solution calorimetry: correlations with stability. J Pharm Sci. 67: 767-772.

[5] Freire, E. (1995). Thermal denaturation methods in the study of protein folding. In *Methods in Enzymology*, Vol. 259 (Johnson, M. L. and Ackers, G. K., Eds.) Academic Press, San Diego, New York, Boston, London, Sydney, Tokyo, Toronto, pp. 144-168.

[6] Takano K, Yamagata Y, Kubota M, Funahashi J, Fujii S, Yutani K 1999. Contribution of hydrogen bonds to the conformational stability of human lysozyme: Calorimetry and x-ray analysis of six ser →ala mutants. Biochem. 38: 6623-6629.

[7] Maneri L R, Farid A R, Smialkowski P J, Seaman M B, Baldoni J M, Sokoloski T D 1991. Preformulation of proteins using high sensitivity differential scanning calorimetry (DSC). Pharm Res. 8: S-48.

[8] Boye J I, Alli I, Ismail A A 1997. Use of differential scanning calorimetry and infrared spectroscopy in the study of thermal and structural stability of α-lactalbumin. J Agric Food Chem. 45: 1116-1125.

---

* Corresponding Author

[9] Daniel RM 1996. The upper limits of enzyme thermal stability. Enzyme Microb. Technol. 19: 74-79.

[10] Quinn ÉÁ 2000. The stability of proteins in hydrofluoroalkane propellants. Ph. D. Thesis, University of Bradford, UK.

[11] Clas S.-D, Dalton CR, Hancock BC 1999. Differential scanning calorimetry: applications in drug development. PSTT 2: 311-320.

[12] Cooper A, Johnson CM (1994). Introduction to microcalorimetry and biomolecular energetics. In Methods in Molecular Biology: Microscopy, Optical Spectroscopy, and Macroscopic Techniques, Vol. 22 (Jones, C., Mulloy, B. and Thomas, A. H., Eds.) Humana Press Inc., Totowa, NJ, pp. 109-124.

[13] Elkordy A A, Forbes R T, Barry B W 2004. Stability of crystallised and spray-dried lysozyme. Int J Pharm. 278: 209-219.

[14] Remmele R L, Nightlinger N S, Srinivasan S, Gombotz W R 1998. Interleukin-1 receptor (IL-1R) liquid formulation development using differential scanning calorimetry. Pharm Res. 15: 200-208.

[15] Branchu S, Forbes R T, Nyqvist H, York P 1998. The relationship between the folding reversibility and enzymatic activity of trypsin. J Pharm Sci. 1 (suppl.): 541.

[16] Forbes R T, Barry B W, Elkordy AA 2007. Preparation and characterisation of spray-dried and crystallised trypsin: FT-Raman study to detect protein denaturation after thermal stress. European J Pharm Sci. 30: 315-323.

[17] Elkordy A A, Forbes R T, Barry B W 2008. Study of protein conformational stability and integrity using calorimetry and FT-Raman spectroscopy correlated with enzymatic activity. European J Pharm Sci. 33: 177-190.

[18] Creighton T E 1994. The protein folding problem. In: Pain, R.H. (Ed.),. Mechanisms of protein folding. Oxford University Press, Oxford, New York, Tokyo, pp. 1-25.

[19] Schmid F X 1992. Kinetics of unfolding and refolding of single-domain proteins. In: Creighton, T.E. (Ed.), Protein folding. Freeman, W.H. and Company, New York, pp. 197-241.

[20] Catanzano F, Giancola C, Graziano G, Barone G 1996. Temperature-induced denaturation of ribonuclease S: A thermodynamic study. Biochem. 35: 13378-13385.

[21] Hirai M, Arai S, Iwase H 1999. Complementary analysis of thermal transition multiplicity of hen egg-white lysozyme at low pH using X-ray scattering and scanning calorimetry. J Phys Chem B. 103: 549-556.

[22] Anfinsen C B 1973. Principles that govern the folding of protein chains. Sci. 181: 223-230.

[23] Katakam M, Bell L N, Banaga A K 1995. Effect of surfactants on the physical stability of recombinant human growth hormone. J Pharm Sci. 84: 713-716.

[24] Elkordy A A, Forbes R T, Barry B W 2002. Integrity of crystalline lysozyme exceeds that of a spray dried form. Int J Pharm. 247: 79-90.

[25] Matouschek A, Serrano L, Fersht A R 1994. Analysis of protein folding by protein engineering. In: Pain, R.H. (Ed.), Mechanisms of protein folding. Oxford University Press, Oxford, New York, Tokyo, pp. 137-159.

[26] Radford S E, Dobson C M, Evans P A 1992. The folding of hen lysozyme involves partially structured intermediates and multiple pathways. Nature. 358: 302-307.

[27] Buck M, Radford S E, Dobson C M 1993. A partially folded state of hen egg white lysozyme in trifluoroethanol: Structural characterization and implications for protein folding. Biochem. 32: 669-678.

[28] Creighton T E 1993. Proteins in solution and in membranes. In: Creighton, T. E. (Ed.),. Proteins: Structures and molecular properties, 2nd Ed., Freeman, W.H. and Company, New York, pp. 287-323.

[29] Lepock J R, Ritchie K P, Kolios M C, Rodahl A M, Heinz K A, Kruuv J 1992. Influence of transition rates and scan rate on kinetic simulations of differential scanning calorimetry profiles of reversible and irreversible protein denaturation. Biochem. 31: 12706-12712.

[30] Koshiyama I, Hamano M, Fukushima D 1981. A heat denaturation study of the 11 S globulin in soybean seeds. Food Chem. 6: 309-322.

[31] Urabe H, Sugawara Y, Ataka M, Rupprecht A 1998. Low-frequency Raman spectra of lysozyme crystals and oriented DNA films: dynamics of crystal water. Biophys J. 74: 1533-1540.

[32] Jen A, Merkle H P 2001. Diamonds in the rough: Protein crystals from a formulation perspective. Pharm Res. 18: 1483-1488.

Permissions

The contributors of this book come from diverse backgrounds, making this book a truly international effort. This book will bring forth new frontiers with its revolutionizing research information and detailed analysis of the nascent developments around the world.

We would like to thank Dr. Amal Ali Elkordy, for lending her expertise to make the book truly unique. She has played a crucial role in the development of this book. Without her invaluable contribution this book wouldn't have been possible. She has made vital efforts to compile up to date information on the varied aspects of this subject to make this book a valuable addition to the collection of many professionals and students.

This book was conceptualized with the vision of imparting up-to-date information and advanced data in this field. To ensure the same, a matchless editorial board was set up. Every individual on the board went through rigorous rounds of assessment to prove their worth. After which they invested a large part of their time researching and compiling the most relevant data for our readers. Conferences and sessions were held from time to time between the editorial board and the contributing authors to present the data in the most comprehensible form. The editorial team has worked tirelessly to provide valuable and valid information to help people across the globe.

Every chapter published in this book has been scrutinized by our experts. Their significance has been extensively debated. The topics covered herein carry significant findings which will fuel the growth of the discipline. They may even be implemented as practical applications or may be referred to as a beginning point for another development. Chapters in this book were first published by InTech; hereby published with permission under the Creative Commons Attribution License or equivalent.

The editorial board has been involved in producing this book since its inception. They have spent rigorous hours researching and exploring the diverse topics which have resulted in the successful publishing of this book. They have passed on their knowledge of decades through this book. To expedite this challenging task, the publisher supported the team at every step. A small team of assistant editors was also appointed to further simplify the editing procedure and attain best results for the readers.

Our editorial team has been hand-picked from every corner of the world. Their multi-ethnicity adds dynamic inputs to the discussions which result in innovative outcomes. These outcomes are then further discussed with the researchers and contributors who give their valuable feedback and opinion regarding the same. The feedback is then collaborated with the researches and they are edited in a comprehensive manner to aid the understanding of the subject.

Apart from the editorial board, the designing team has also invested a significant amount of their time in understanding the subject and creating the most relevant covers. They scrutinized every image to scout for the most suitable representation of the subject and create an appropriate cover for the book.

The publishing team has been involved in this book since its early stages. They were actively engaged in every process, be it collecting the data, connecting with the contributors or procuring relevant information. The team has been an ardent support to the editorial, designing and production team. Their endless efforts to recruit the best for this project, has resulted in the accomplishment of this book. They are a veteran in the field of academics and their pool of knowledge is as vast as their experience in printing. Their expertise and guidance has proved useful at every step. Their uncompromising quality standards have made this book an exceptional effort. Their encouragement from time to time has been an inspiration for everyone.

The publisher and the editorial board hope that this book will prove to be a valuable piece of knowledge for researchers, students, practitioners and scholars across the globe.

List of Contributors

Safia Alleg and Saida Souilah
Badji Mokhtar Annaba University, Department of Physics, Laboratoire de Magnétisme et Spectroscopie des Solides (LM2S) B.P. 12, 23000 Annaba, Algeria

Joan Joseph Suñol
Dep. De Fisica, Universitat de Girona, Campus Montilivi, 17071 Girona, Spain

Adriana Gregorova
Graz University of Technology, Institute for Chemistry and Technology of Materials, Austria

P.V. Dhanaraj
Department of Physics, Malabar Christian College, Kozhikode, India

N.P. Rajesh
Centre for Crystal Growth, SSN College of Engineering, Kalavakkam, India

Jose C. Martinez, Javier Murciano-Calles, Eva S. Cobos, Manuel Iglesias-Bexiga, Irene Luque and Javier Ruiz-Sanz
Department of Physical Chemistry and Institute of Biotechnology, Faculty of Sciences, University of Granada, Granada, Spain

Ruel E. McKnight
Department of Chemistry, State University of New York at Geneseo, 1 College Circle, Geneseo, NY, USA

Diana Romanini, Mauricio Javier Braia and María Cecilia Porfiri
Laboratory of Physical Chemistry Applied to Bioseparation. College of Biochemical and Pharmaceutical Sciences, National University of Rosario (UNR), Rosario, Argentina

Stefka G. Taneva
Unidad de Biofísica (CSIC/UPV-EHU), Departamento de Bioquímica y Biología Molecular, Universidad del País Vasco, Bilbao, Spain Institute of Biophysics and Biomedical Engineering, Bulgarian Academy of Sciences, Sofia, Bulgaria

Laura T. Rodriguez Furlán, Antonio Pérez Padilla and Mercedes E. Campderrós
Research Institute of Chemical Technology (INTEQUI –CONICET-CCT San Luis), Faculty of Chemistry, Biochemistry and Pharmacy, UNSL, San Luis, Argentine

Javier Lecot and Noemi E. Zaritzky
Research and Development in Food Cryotechnology Centre (CIDCA- CONICET- CCT La Plata), Argentine

Noemi E. Zaritzky
Faculty of Engineering, UNLP, La Plata, Bs As, Argentine

Amal A. Elkordy
Department of Pharmacy, Health and Well-being, University of Sunderland, Sunderland, UK

Robert T. Forbes and Brian W. Barry
School of Pharmacy, University of Bradford, Bradford, UK

Printed in the USA
CPSIA information can be obtained
at www.ICGtesting.com
JSHW011417221024
72173JS00004B/569